Robust Battery Management System Design with MATLAB®

For a complete listing of titles in the
Artech House Power Engineering Series,
turn to the back of this book.

Robust Battery Management System Design with MATLAB®

Balakumar Balasingam

ARTECH
HOUSE

BOSTON | LONDON
artechhouse.com

Library of Congress Cataloging-in-Publication Data
A catalog record for this book is available from the U.S. Library of Congress.

British Library Cataloguing in Publication Data
A catalogue record for this book is available from the British Library.

Cover design by Andy Meaden

ISBN 13: 978-1-63081-952-1

Accompanying software to this book can be found at:
https://github.com/SingamLabs/Robust-Battery-Management-Systems.git

© **2023 ARTECH HOUSE**
685 Canton Street
Norwood, MA 02062

10 9 8 7 6 5 4 3 2 1

Contents

Preface

Rechargeable batteries are becoming ubiquitous, and battery management has become a topic of increasing interest for system designers, engineers, research scientists, professors, and graduate students. There is a need to continuously study and improve battery management systems so that the safety, efficiency, and reliability of battery power systems can be further improved. Particularly, there is a need to introduce battery management as an evolving field to graduate students and research scientists so that it can continue to improve through additional research. At the same time, there is a need to present practical solutions that are well founded in theory and easily explainable to engineers and system designers to adequately solve various battery management problems based on the credible progress made so far; this book tries to achieve these two objectives. Particularly, the following battery management problems are covered in detail: battery impedance estimation, battery capacity estimation, state of charge estimation, state of health estimation, battery thermal management, optimal charging algorithms. Accompanied MATLAB® codes give an easy way for the readers to test the algorithms using realistic data and to develop and test alternative solutions on their own.

The insight and necessary material for this book were developed based on various research activities that I undertook over the past 12 years. I was first introduced to the problem of battery management in late 2011 by Professor Krishna Pattipati while I was a postdoctoral fellow at the University of Connecticut. By working with Professor Pattipati on solving various battery management problems, I firsthand witnessed and learned how one can apply theoretical skills to solve practical engineering problems, and I was hooked. This solid foundation led me to become a professor and develop a research lab that is focused on improving the state of the art in battery management systems at the University of Windsor. I am now able to confidently guide graduate students on pushing the boundaries of battery management. Many of the results presented in this book are the results of the enthusiasm and hard work of my graduate students at the University of Windsor. Particularly, I am thankful to Prarthana Pillai, PhD student, who helped me with this book from the first chapter to the last by providing all aspects of technical skills needed to make this book a reality.

Chapter 1

About This Book

1.1 INTRODUCTION

Rechargeable lithium (Li) ion batteries are used to power many things: consumer electronics, power equipment, household appliances, aerospace systems, and electric vehicles are a few examples. A battery management system (BMS) is essential for any rechargeable battery to ensure safety, efficiency, and reliability. Due to the growth in electric vehicles, interest in battery management systems has grown steadily over the past decade and is expected to grow significantly in the foreseeable future. Battery management systems and related topics have started to appear in university curriculums. Indeed, this book grew out of a graduate course on battery management systems.

Electric vehicle batteries are made of thousands of battery cells connected in series and in parallel; cells are connected in series to increase the voltage output, and they are connected in parallel to increase the battery capacity. Each battery cell within a battery pack needs to be constantly monitored to ensure that it is kept within allowable limits. Battery management systems are a very crucial part of electric vehicles, numerous other devices, and energy storage systems that rely on battery packs for energy storage.

This book introduces various aspects of battery management systems. Particularly, this book adopts the equivalent circuit model approach to battery management and explains various elements of it. This book explains well-known solutions strategies as well as novel approaches that are based on my research work.

1.2 WHO IS THIS BOOK FOR?

Electric vehicles (EVs) are about to replace internal combustion engines due to wider adaptation expected in the near future. Internal combustions engines were adopted

1

more than a century ago; to this day, research and development aimed at improving the performance of internal combustion engine continues. A similar trajectory can be expected of battery electric vehicles (BEVs), which started to see early adoption quite recently. Battery engineer is a leading job title in automobile manufacturing companies. Battery engineers are also in demand in other industries, such as consumer electronics, household equipment, power tools, Internet of Things, aerospace, and defense. This book is a useful handbook to battery engineers seeking to teach themselves about the basics of battery management.

Today's battery packs are heavier than internal combustion engines; the range of present day BEVs is very low and their charging time remains a bottleneck for wider BEV adoption; and the purchase cost of BEVs is significantly higher than their traditional counterparts. This indicates the need for continued research about improvements of battery management systems. To help improve the BEV technology, there is significant research interest about batteries in government and private research laboratories. This book will be a useful tool to those researchers working to improve the state-of-the-art in BEVs.

Researchers about new battery chemistries aim to develop better battery chemistries that can last longer, charge faster, cost less, weigh less, and store and discharge higher amounts of electrons. However, the adoption of newer battery chemistries into BEVs depend on numerous constraints involving environmental, economic, and engineering concerns. A BEV battery pack could outlive its primary application due to its potential use in second-use applications. The fundamental problems and solutions concerning battery management as we know them today will likely remain relevant for years to come. This book is hence useful as a textbook to university professors and researchers interested in developing a course on battery management systems. Finally, this book will be useful to undergraduate and graduate-level students interested in electrified transportation, autonomous systems, robotics, Internet of Things, and many other relevant fields requiring knowledge about batteries and battery management.

1.3   USE CASES

This book may be used as a reference to established battery management algorithms that are based on the electrical equivalent circuit models. The book tries to clearly define different aspects of the battery management problems and describes solutions. In this section, we explain some use cases for this book.

1.3.1 Remaining Mileage Estimation in an Electric Vehicle

Displaying the remaining available mileage is crucial to BEVs. The time to shutdown (TTS) formula defined in Chapter 8 is a generic form to compute the remaining time until the battery needs to shutdown for recharging. The TTS computation requires the state of charge (SOC) and battery impedance information that can be computed based on the algorithms provided in Chapters 8 and 6, respectively. Recent studies about state-of-the-art remaining mileage indicators in vehicles suggest that there is significant error in these displays. The solution to this problem involves deeper understanding of battery characterization, state of charge estimation, and ECM identification. It is also important to understand the implications of sensor bias and measurement noise to enhance the performance of the state-of-the-art solutions. The book provides the necessary background to understand, implement, and improve battery fuel gauging.

1.3.2 Generating Battery Replacement Warning

The available mileage of BEVs reduces with time, depending on the charging and usage patterns, environmental factors, and battery age. Battery management systems need to give an accurate estimate of the state of health of the battery. This book provides details about computing power and capacity fade and using that to estimate the state of health of a battery. System designers can easily incorporate the MATLAB® provided in this book to test the performance of several algorithms presented in this book against alternative methods. All the algorithms presented in this book can be tested by simulating battery data using the MATLAB-based battery simulator provided in this book.

1.3.3 Estimating the Expected Temperature Rise in a Battery Pack

Battery thermal management is an important topic in order to ensure the safety and efficiency of an electric vehicle. Research shows that maintaining the battery within specific temperature levels ensures safety and extends its life span. Existing thermal management systems are reactive (i.e., they sense the temperature increase and then respond); this book teaches the necessary tools to develop a predictive cooling system that is more efficient (and more driving miles) than a reactive thermal management system. Proper thermal management will enhance the safety and reliability of a battery pack, provide more driving miles, and slow down the state of health degradation. The book provides models and algorithms to estimate the thermal properties of a battery.

1.3.4   Smart Battery Charger Design

A battery pack made for a specific application must be accompanied by a charger that is designed to charge that specific type of battery pack. For this specific charger, a battery engineer needs to finalize the parameters of the constant-current constant-voltage (CCCV) charger. Chapter 10 details how the parameters of the CCCV charger can be selected in a way that both the charging time and energy loss can be optimized. Other chapters in this book detail how required parameters for the smart charger, such as internal resistance and battery capacity, can be continuously estimated to design a charger that adapts with the age and state of health of the battery.

1.3.5   EV Fleet Management

An EV fleet manager tries to optimize the cost that is measured in terms of the required number of EVs, the cost of charging them, operational delays, and maintenance costs of batteries. The constraints may include state of charge, state of health, required charging time, and remaining useful life of a battery, quantities that can be computed based on the algorithms presented in this book.

1.3.6   Teaching a Graduate-Level Course on BMS

As the world transitions from internal combustion engines to electric vehicles, the battery pack, which is a relatively new phenomenon compared to electric motors and power transmission systems, attracts the attention of engineers, chief executive officers, and government officials. Battery packs of the present have plenty of room for improvement; this requires widespread studies about how to improve the performance of the present-day battery packs and how to use the lessons learned today to make better battery packs of the future. In order to motivate innovation in this domain, several elements of a battery management system can to be taught in universities to the next generations of engineers and scientists. Indeed, this book grew out of a graduate course on battery management systems in the electrical engineering discipline.

1.4   WHAT IS NOVEL IN THIS BOOK?

Battery management is an active research area. Numerous research papers are being published on the subject in present-day literature. Some notable previous books on this topic are [1–5]. This book contributes to the literature in the following ways.

1.4.1   Modularized Approach

The book introduces important battery management problems in a modularized fashion; the overall battery management problem is divided into smaller modules, such as open-circuit voltage (OCV) characterization, state of charge estimation, and equivalent circuit model identification. Dividing battery management into smaller modules or blocks allows one to focus on a particular module without having to worry about BMS at its entire system level. Each chapter in this book starting from Chapter 4 is dedicated to explaining one particular aspect of a battery management system.

1.4.2   Illustration of Algorithms Through MATLAB Simulation

MATLAB simulations offer an easy and effective way to understand the problems and test solutions through algorithms presented in the book. In order to test algorithms, voltage and current data emulating a battery can be obtained using a battery simulator provided within this book. Several MATLAB examples explaining how to use the battery simulator are also provided. Working on algorithms with simulated data, rather than real voltage, current data collected from a battery, allows beginners to objectively understand the effect of each modules comprising a battery management system.

1.4.3   Emphasis on Both Theoretical and Practical Aspects

This book is written to introduce various battery management problems and to offer guidance on solution development. Effective battery management system development requires both theoretical development and practical insights; this book aims to stimulate both theoretical investigation and practical view of problems. On the theory side, this book either derives or alludes to the existence of theoretical performance bounds of various state and parameter estimation algorithms. These discussions are aimed at researchers and scientists working on improving the state-of-the-art in battery management algorithms. On practical aspects, the book emphasizes on the implementation aspects of algorithms and on practical ways to evaluate the performance of battery management algorithms. Chapter 11 discusses ways to evaluate the performance of a battery management system in real-world settings. Continued emphasis on measurement noise and its effect on the performance is discussed throughout the book. Many algorithms, available from the current literature, work in low noise environments; their performance may significantly degrade in the presence of measurement noise, which is abundant in low-cost (and hence preferred) sensors and measurement devices. Redeveloping many of the battery management strategies to highly noisy environments is an ongoing effort; this book offers insights into this very practical aspect.

1.5   ORGANIZATION OF THIS BOOK

This book is organized into 11 chapters. This chapter provides an introduction to the book. An overview of what is covered in each of the remaining chapter is presented below.

- Chapter 2: Review of Required Mathematics. Battery management systems, in an algorithmic perspective, need significant insights into system engineering. Hence, an undergraduate level of knowledge in system engineering is a prerequisite for this book. A good background in matrix linear algebra [6] and probability theory [7] is beneficial to follow the details of algorithms presented in this book. Strong graduate-level knowledge in estimation theory [8] is also beneficial to follow the details in some chapters. Chapter 2 briefly summarizes some required mathematical concepts in estimation theory that are later utilized in various chapters of this book.

- Chapter 3: Battery Modeling. In order to develop advanced battery management functionalities, a mathematical model of the battery is needed. A battery model can be considered as the theoretical or hypothetical twin of a battery or battery pack; it is built based on both theoretical derivations and empirical modeling. Most battery management is done based on an assumed underlying battery model. When the model assumption is different from reality, the functionality of a BMS becomes less accurate. This chapter presents the details of electrical equivalent circuit model of a battery, consisting of the following components: an open-circuit voltage or electromotive force model, models to represent the hysteresis effect, and models that represent the relaxation effect of the battery.

- Chapter 4: Open-Circuit Voltage Characterization. The open-circuit voltage is an important phenomenon in battery management. The measured voltage across battery terminals changes even when there is no current through it; this is due to the relaxation effect within the battery. A battery is considered to be at rest after it experiences zero current for several hours. When the battery is rested, the measured voltage across its terminals is equal to the open-circuit voltage. The open circuit voltage of a battery has a monotonous relationship with its SOC. The OCV-SOC characterization is useful for many battery management functionalities, such as, state of charge and state of health estimation. The OCV-SOC relationship can be mathematically established based on electrochemical principles; however, it requires precise information about the chemical compositions of the battery, which may change as the battery ages. Instead, the state-of-the-art in battery management is to obtain the OCV-SOC characterization through empirical means. This chapter provides details about OCV-SOC modeling.

- Chapter 5: Frequency-Domain Approaches to Battery ECM Identification. The relaxation effect of the battery is modeled using electrical components, namely, resistors, capacitors, and inductors. Such a model forms part of the equivalent circuit model (ECM) of a battery. The ECM parameters can be identified and estimated through both frequency-domain and time-domain means. In frequency-domain ECM analysis, voltage/current excitation signals at various frequencies are applied to the battery and the response current/voltage is measured to obtain the impedance spectrum, also referred to as the Nyquist spectrum, of the battery. By analyzing the impedance spectrum, ECM parameters can be obtained. This chapter details frequency-domain approaches to battery ECM identification and parameter estimation. This chapter also highlights two limitations of the frequency-domain approach (that they are sensitive to measurement noise and that they are time-consuming) and offers insights about improvements.

- Chapter 6: Time-Domain Approaches to Battery ECM Identification. An attractive feature of the frequency-domain approach to ECM identification is that some of the ECM parameters can be visually identified from the impedance spectrum; hence, it was adopted by a wide range of research communities for impedance analysis in general and battery analysis in particular. However, frequency-domain approaches require special hardware for measurements and are time consuming. Time-domain approaches use opportunistic measurement, without the requirement of any special hardware, for ECM identification. Various time-domain approaches to ECM parameter estimation are covered in this chapter. The effect of measurement noise is formally defined and approaches are discussed for efficient ECM parameter estimation in the presence of measurement noise.

- Chapter 7: Battery Capacity Estimation. The ability to store and release electrons in a battery decreases over time; this phenomenon is known as capacity fade. The rate of capacity fade is affected by various factors, such as charge/discharge rate, temperature, humidity, depth of discharge, and calendar age. Collecting and tracking all these variables are very challenging tasks; instead, capacity fade is quantified by estimating the battery capacity in real time. Two different approaches to real-time capacity estimation of a battery are discussed in this chapter: one exploits battery resting instances and the other makes use of the time-domain ECM parameter identification technique described in Chapter 6. The pros and cons of these two approaches are quantitatively discussed. Once again, this chapter highlights the perils of measurement noise and discusses ways to quantify them and to assess the performance of the approaches by considering the amount of measurement noise.

- Chapter 8: Battery Fuel Gauging. Battery fuel gauging is thus far the most researched, discussed, and published topic in the battery management literature. Output of a battery fuel gauge includes the state of charge, time to shutdown (or remaining mileage of an EV), state of health, and remaining useful life. Numerous approaches have been reported to estimate the state of charge of a battery. This chapter details three important approaches for state of charge estimation: one based on Coulomb counting, another based on voltage lookup, and the other based on fusing both Coulomb counting and voltage-based approaches; the extended Kalman filter is widely employed in the fusion-based approach. Some practical details about how to monitor the extended Kalman filter for possible failures during SOC estimation and how to make it robust are also discussed.

- Chapter 9: Battery Thermal Management. Battery thermal management is a vast topic, especially when it comes to EV battery pack design. Many aspects of battery thermal management, selection of heating/cooling topology, selection of material for battery pack design, placement of cells, packs, and modules are important for battery thermal management, but are not considered in this book. For any battery thermal management system design, the amount of expected heat and the temperature rise during regular operations, such as charging and driving in the case of EVs, is an important consideration. This chapter describes a thermal-electrical equivalent circuit model approach to compute the expected temperature rise in a battery.

- Chapter 10: Optimal Charging Algorithms. A battery charger needs to reduce three competing features: charging time, heat dissipation, and state of health degradation. Optimized battery charging remains an active research field and there are numerous research works aimed at finding better ways to fast charge a battery. Li-ion battery charging is a challenging problem due to several constraints. The battery terminal voltage cannot exceed certain respective thresholds due to safety constraints; this results in extended charging time. High charging current causes energy loss that results in temperature rise; high charging current is also suspected to be a cause of battery degradation. The charging current must be controlled in a way that energy loss is minimal, the charging time is acceptable, and the state of health of the battery is not degraded at unwanted levels. This chapter introduces various battery charging methods and shows how competing objectives can be met by formulating and solving optimization problems.

- Chapter 11: Evaluation and Benchmarking of Battery Management Systems. This chapter introduces the challenges involved in evaluating a battery management algorithm. It then details three approaches to evaluate the performance of a battery fuel gauge.

1.6 MATLAB CODES

The subsequent chapters of this book contains several MATLAB examples. Some MATLAB codes are printed in the book; some codes are made available in the companion GitHub website that can be found here: https://github.com/SingamLabs/Robust-Battery-Management-Systems.git. Some codes, with some extra descriptions, are also made available through specific repositories in Mendeley; these repositories are cited in the respective chapters.

1.7 BIBLIOGRAPHICAL NOTES

In [1], system-level descriptions and high-level introduction to battery management systems are presented with a particular focus on practical and implementation aspects. Various components involved in a battery management system development and their functionalities were presented in [2]. In [3], practical aspects of various battery management system functionalities were presented. In-depth details about the basics of Li-ion batteries, battery modeling, and battery management approaches can be found in a two-part book series [4, 5]; the DC equivalent circuit models of a battery presented in this book are adopted from it. This book assumes graduate-level knowledge of linear algebra [6], probability [7], and estimation theory [8]. Some of the required mathematical concepts are briefly reviewed in Chapter 2; for additional details about any mathematical concepts used in this book, the readers are referred to [6–8].

References

[1] D. Andrea, *Battery Management Systems for Large Lithium Ion Battery Packs,* Artech House, Norwood, MA, 2010.

[2] P. Weicker, *A Systems Approach to Lithium-Ion Battery Management,* Artech House, Norwood, MA, 2010.

[3] Y. Barsukov, and J. Qian, *Battery Power Management for Portable Devices,* Artech House, Norwood, MA, 2013.

[4] G.L. Plett, *Battery Management Systems, Volume I: Battery Modeling,* Artech House, Norwood, MA, 2015.

[5] G.L. Plett, *Battery Management Systems, Volume II: Equivalent-Circuit Methods*, Artech House, Norwood, MA, 2015.

[6] G. Strang, *Introduction to Linear Algebra,* Cambridge University Press, Cambridge, UK, 2016.

[7] J.A. Gubner, *Probability and Random Processes for Electrical and Computer Engineers,* Cambridge University Press, Cambridge, UK, 2006.

[8] Y. Bar-Shalom, X.R. Li, and T. Kirubarajan, *Estimation with Applications to Tracking and Navigation: Theory Algorithms and Software,* John Wiley & Sons, New York, 2004.

Chapter 2

Review of Required Mathematics

2.1 INTRODUCTION

Battery management systems are designed based on system theory. Undergraduate-level topics on probability, matrix linear algebra, signals and systems, and numerical methods are useful to understand and build upon the concepts presented in this book. At the graduate level, this book requires sound knowledge of estimation theory. This chapter provides a brief review of estimation theory that is used later in subsequent chapters. For more detailed review of estimation theory, the reader is referred to [1].

Several aspects of battery management systems require the estimation of various battery states and parameters. Parameters are those that do not change with time and states change over time. From a battery perspective, parameters include resistors, capacitors, and inductors that form the equivalent circuit model of a battery; examples of states are state of charge (SOC) and state of health (SOH). This chapter gives a brief introduction to estimation theory covering both deterministic (least square) and probabilistic (Bayesian) estimation concepts.

This chapter assumes undergraduate-level knowledge of matrix linear algebra and probability theory. Whenever necessary, the reader is directed to [2] and [3] to refer to topics related to linear algebra and probability theory, respectively.

2.2 LEAST SQUARES ESTIMATOR

Let us consider the following observation model

$$z = Hx + w \tag{2.1}$$

where the $m \times 1$ vector $\mathbf{z}$ is a measurement about the $n \times 1$ vector $\mathbf{x}$, the matrix $\mathbf{H} \in \mathfrak{R}^{m \times n}$ is the observation matrix and $\mathbf{w} \in \mathfrak{R}^m$ represents the measurement noise. Least-squares (LS) estimation seeks to find $\hat{\mathbf{x}}_{LS}$ such that the square error is minimized, that is,

$$\hat{\mathbf{x}}_{LS} = \arg \min_{\mathbf{x}} \|\mathbf{H}\mathbf{x} - \mathbf{z}\|_2^2 \tag{2.2}$$

where the observation matrix $\mathbf{H}$ is assumed known, tall ($m > n$), and full rank. In determining $\hat{\mathbf{x}}_{LS}$, we find that value of $\mathbf{x}$ which provides the best fit of the observations to the model in the 2-norm sense. The expression $\|\mathbf{H}\mathbf{x} - \mathbf{z}\|_2^2$ can be written as

$$\|\mathbf{H}\mathbf{x} - \mathbf{z}\|_2^2 = (\mathbf{H}\mathbf{x} - \mathbf{z})^T (\mathbf{H}\mathbf{x} - \mathbf{z}) \tag{2.3}$$
$$= \mathbf{z}^T\mathbf{z} - \mathbf{x}^T\mathbf{H}^T\mathbf{z} - \mathbf{z}^T\mathbf{H}\mathbf{x} + \mathbf{x}^T\mathbf{H}^T\mathbf{H}\mathbf{x} \tag{2.4}$$

The least-square estimate $\hat{\mathbf{x}}_{LS}$ is that value of $\mathbf{x}$ that satisfies

$$\frac{d}{d\mathbf{x}} \left[\mathbf{z}^T\mathbf{z} - \mathbf{x}^T\mathbf{H}^T\mathbf{z} - \mathbf{z}^T\mathbf{H}\mathbf{x} + \mathbf{x}^T\mathbf{H}^T\mathbf{H}\mathbf{x} \right] = 0 \tag{2.5}$$

It can be shown that the above differentiation reduces to the following form [4]

$$\mathbf{H}^T\mathbf{H}\mathbf{x} = \mathbf{H}^T\mathbf{z} \tag{2.6}$$

From the above, the LS estimate of the vector $\mathbf{x}$ is obtained as

$$\hat{\mathbf{x}}_{LS} = \left(\mathbf{H}^T\mathbf{H}\right)^{-1} \mathbf{H}^T\mathbf{z} \tag{2.7}$$

The LS estimator is subject to the following assumptions

1. Linearity observation model: The observation model in (2.1) is linear.
2. Additive noise: The noise is additive; this is reflected in the cost function (2.2). The following is an example of nonadditive noise, particularly multiplicative noise.

$$\mathbf{z} = \mathbf{H}\mathbf{x} \odot \mathbf{w} \tag{2.8}$$

where $\odot$ is the element-by-element multiplication operator.
3. Known model: The matrix $\mathbf{H}$ in (2.1) is known as the observation matrix. The LS estimate assumes that the observation matrix is completely known.

Example 2.1 Resistance Estimation

Figure 2.1 shows a measurement setup designed to estimate the resistance of an unknown resistor R. The resistor is connected to a current source that provides a constant current of 1 A. While the resistor is connected to the current source, several voltage measurements across the resistor were taken using a volt meter, shown in the diagram. Develop a least-squares approach to estimate the resistance R.

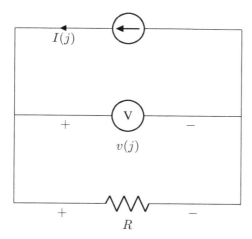

Figure 2.1 Measurement setup. A resistor is connected to a constant current source that continuously provides $I(j) = 1$ A. The voltage across the resistor is measured by a voltmeter at fixed sampling intervals and these measured voltages are denoted as $v(j)$, $k = 1, 2, \ldots, m$. The goal is to estimate the value of the resistance R using the least-squares approach.

The measured voltage across the resistor for a given current can be written as

$$v(j) = I(j)R + n(j) \tag{2.9}$$

where $v(j)$ is the measured voltage at time j, $I(j) = I_c = 1$A is the current, which is assumed known and constant, at time j, and $n(j)$ is the corresponding measurement noise at time j. Considering all m consecutive observations, the above observation can be written in vector form as

$$\mathbf{v} = \mathbf{I}R + \mathbf{w} \qquad (2.10)$$

where $\mathbf{v}, \mathbf{I}$, and $\mathbf{w}$ are vectors of length m, that is,

$$\mathbf{v} = \begin{bmatrix} v(1) \\ v(2) \\ \vdots \\ v(m) \end{bmatrix}, \ \mathbf{I} = \begin{bmatrix} I_c \\ I_c \\ \vdots \\ I_c \end{bmatrix}, \ \mathbf{w} = \begin{bmatrix} w(1) \\ w(2) \\ \vdots \\ w(m) \end{bmatrix} \qquad (2.11)$$

Now it is easy to see that the observation model (2.9) has the same form as (2.8). The least-square estimate of R is then given by

$$\hat{R}_{\mathrm{LS}} = \frac{\mathbf{I}^T \mathbf{v}}{\mathbf{I}^T \mathbf{I}} \qquad (2.12)$$

Table 2.1 shows 10 different voltage measurements $v(j)$ for $j = 1, 2, \ldots, 10$ corresponding to Example 2.1. One would expect the measured voltage to be constant because the current through the resistor is set to be a constant $I(j) = I_c = 1\mathrm{A}$ and the resistance itself is a constant. Hence, the variances in the measured voltage are attributed to the measurement noise; even though the individual value of each measurement noise $w(j)$ is not known, their presence is found based on the variance in the observed voltage measurements.

    If one were to ignore the presence of measurement noise to estimate the resistance, the resistance estimate can be written as $\hat{R} = v(j)/I(j)$. Based on this approach, the estimated resistance would be different based on each voltage measurement. Table 2.1 lists the estimated resistance values corresponding to the voltage measurements across the resistor R. Using (2.12) and the data from Table 2.1, the LS estimate of R in Figure 2.1 is obtained as

$$\hat{R}_{\mathrm{LS}} = 0.2062 \, \Omega \qquad (2.13)$$

After it is given that the true resistance is $R = 0.2 \, \Omega$, one can see that the LS estimate is much superior to the individual estimates reported in Table 2.1. □

Table 2.1

Voltage Measurements and Individual Estimates Across R

j	$v(j)$	$\hat{R}\,(\Omega) = v(j)/I(j)$
1	0.2054	0.2054
2	0.2183	0.2183
3	0.1774	0.1774
4	0.2086	0.2086
5	0.2032	0.2032
6	0.1869	0.1869
7	0.1957	0.1957
8	0.2034	0.2034
9	0.2358	0.2358
10	0.2277	0.2277

Example 2.2 EMF and Resistance Estimation

Figure 2.2 shows an equivalent circuit model of a battery. The electromotive force of the battery (in volts) is denoted as EMF and the internal resistance of the battery is denoted as R_0. In order to estimate the internal resistance, a constant current of $i(k) = 2$A is applied to the battery and the resulting terminal voltage $v(k)$ is measured 10 consecutive times. The measured voltages $v(1)$, $v(2)$, ..., $v(10)$, are given by 4.2057, 4.2154, 4.2041, 4.2024, 4.2174, 4.1839, 4.1927, 4.2048, 4.1838, and 4.2037, respectively. Then the current is changed to $i(k) = 1$A and the resulting terminal voltage $v(k)$ is measured 10 consecutive times. The measured voltages $v(11)$, $v(12)$, ..., $v(20)$, are given by 4.0057, 3.9980, 3.9967, 3.9898, 3.9947, 4.0055, 4.0038, 3.9894, 4.0041, and 3.9955, respectively. Use the least-square estimation approach to estimate the EMF and internal resistance R_0 of the battery. Assume that the EMF and R_0 remained constant throughout the experiment.

The voltage measurement at each instant k can be written as

$$v(1) = \mathrm{E} + i(1)R_0 + w(1) \qquad (2.14)$$

$$\vdots \quad \vdots \quad \quad \vdots \quad \quad \quad \vdots$$

$$v(m) = \mathrm{E} + i(m)R_0 + w(m) \qquad (2.15)$$

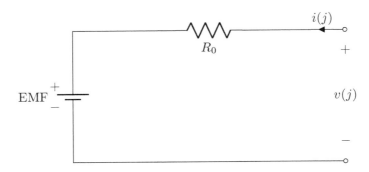

Figure 2.2 Equivalent circuit model of a battery.

where $E \triangleq EMF$ and $m = 20$. In the matrix notation, the equation is thus

$$\mathbf{z} = \mathbf{Hx} + \mathbf{w} \qquad (2.16)$$

where

$$\mathbf{z} = \begin{bmatrix} v(1) \\ v(2) \\ \vdots \\ v(m) \end{bmatrix}, \ \mathbf{H} = \begin{bmatrix} 1 & i(1) \\ 1 & i(2) \\ \vdots & \vdots \\ 1 & i(m) \end{bmatrix}, \ \mathbf{x} = \begin{bmatrix} E \\ R_0 \end{bmatrix}, \ \mathbf{w} = \begin{bmatrix} w(1) \\ w(3) \\ \vdots \\ w(m) \end{bmatrix} \qquad (2.17)$$

For the observation model in (2.16), the least-square estimate $\mathbf{x}$ is given by

$$\hat{\mathbf{x}}_{\text{LS}} = \begin{bmatrix} \hat{E} \\ \hat{R}_0 \end{bmatrix} = (\mathbf{H}^T\mathbf{H})^{-1}\mathbf{H}^T\mathbf{z} = \begin{bmatrix} 3.7952 \\ 0.2031 \end{bmatrix} \qquad (2.18)$$

☐

2.3 KALMAN FILTER

The observation model introduced in Section 2.2 assumes that the unknown variable $\mathbf{x}$ does not change over time (i.e., the variable $\mathbf{x}$ was assumed to be a constant). For example, the resistance R in Example 2.1 is assumed to be a constant; if the resistance were to change with time k, then the least-squares approach may not be suitable to estimate the resistance.

In many practical applications, the parameters of interest change with time. For example, the internal resistance of a battery changes over time due to changes in temperature, state of charge, and state of health of the battery; other components of a battery equivalent circuit model are likely to change with these factors as well. The capacity of the battery also changes over time; several experimental studies showed incremental decay in battery capacity due to aging and many other factors affecting the state of health of the battery. The state of charge of the battery changes instantaneously based on the rate of current. This section focuses on extending the least-squares estimation approach to the state estimation problem where the parameter of interest changes with time k.

When a parameter of interest $\mathbf{x}$ changes over time, we refer to it as a state. By considering equal sampling time, we will denote the state of $\mathbf{x}$ at time k as $\mathbf{x}(k)$. The transition of the state from one time instant (k) to the next $(k + 1)$ is modeled based on the underlying physical phenomenon. For example, let us denote $x(k)$ to be the state of charge of the battery; the state of charge at the next time instant, $x(k+1)$, depends on the current through the battery between the time instant k and $k + 1$ and the battery capacity. More details about the state of charge modeling over time can be found in Chapter 8.

The change of state is a continuous process. For example, consider the resistor in Example 2.1; if the resistance R were to change with time, the change will happen continuously. Let us consider two time instances from time t_k until t_{k+1}. Assuming that the desired state $x = R$ changes over time, there are infinite values of the state x within the time interval spanning t_k and t_{k+1}. Instead of trying to estimate infinite values of x, the state estimation approach presented in this section will estimate the state at discrete points: the state at time t_k is denoted as $x(k)$ and the state at time t_{k+1} is denoted as $x(k + 1)$. The time difference $\Delta_k = t_{k+1} - t_k$ serves to indicate the time difference between $x(k)$ and $x(k+1)$. The evolution of $\mathbf{x}(k)$ over time is captured in the following process model:

$$\mathbf{x}(k) = \mathbf{F}\mathbf{x}(k - 1) + \mathbf{v}(k) \tag{2.19}$$

where the state $\mathbf{x}(k)$ is written in vector format, the matrix $\mathbf{F}$ is known as the state transition matrix, and $\mathbf{v}(k)$ is known as the process noise. The state transition matrix $\mathbf{F}$ is defined based on the physical properties that (approximately) define the change in the state from $\mathbf{x}(k)$ to $\mathbf{x}(k + 1)$. The structure of (2.19) also tells one that the state transition can be approximated as a linear model. In some applications, such linear approximation is not possible. Section 2.4 elaborates more on nonlinear state-space models and estimation approaches. Given that the true state transition is continuous and the discrete-time model is only an approximate representation capturing the change of state from time t_k to time t_{k+1}, the process noise $\mathbf{v}(k)$ accounts for the modeling error.

In many practical applications, this modeling uncertainty is represented as a zero-mean Gaussian noise whose covariance matrix is written as

$$E\left[\mathbf{v}(k)\mathbf{v}(k)^T\right] = \mathbf{Q} \tag{2.20}$$

The process equation (2.19), which is also known as the plant equation in the literature, is written in the most general vector form. Let us assume $\mathbf{x}(k)$ to be an $n \times 1$ vector; then the state transition matrix $\mathbf{F}$ and the process covariance matrix $\mathbf{Q}$ are both $n \times n$ matrices. Similar to before in (2.1), the observation model is written as

$$\mathbf{z}(k) = \mathbf{H}\mathbf{x}(k) + \mathbf{w}(k) \tag{2.21}$$

where, $\mathbf{z}(k)$ is assumed to be an $m \times 1$ observation vector. $\mathbf{H}$ is known as the observation matrix and the measurement noise $\mathbf{w}(k)$ is assumed to be zero-mean Gaussian with covariance defined as

$$E\left[\mathbf{w}(k)\mathbf{w}(k)^T\right] = \mathbf{R} \tag{2.22}$$

which is an $m \times m$ matrix.

Now the state estimation problem can be formally stated as follows: Given the measurements $\mathbf{z}(1)$, $\mathbf{z}(2)$, ..., $\mathbf{z}(L)$ estimate the corresponding states $\mathbf{x}(1)$, $\mathbf{x}(2)$, ..., $\mathbf{x}(L)$ in real-time. Here, even though the state of interest is a continuously varying quantity, its realizations at time instances $t_1, t_2, \ldots, t_L$ are to be estimated based on measurements $\mathbf{z}(t_1)$, $\mathbf{z}(t_2)$, ..., $\mathbf{z}(t_m)$ obtained at those same time instances.

The term real-time estimation needs some clarification; this is done by comparing and contrasting recursive estimation with batch estimation. In real-time estimation, the state at time k is estimated based on the available measurements $\mathbf{z}(1), \mathbf{z}(2), \ldots, \mathbf{z}(k)$ at that time. Table 2.2 lists the available measurements in a real-time state estimation problem. Here, as time k increases, so does the available measurements that can be used to estimate the state $\mathbf{x}(k)$. Table 2.2 also denotes the estimate of $\mathbf{x}(k)$ given the measurements $\mathbf{z}(1), \mathbf{z}(2), \ldots, \mathbf{z}(k)$ as $\hat{\mathbf{x}}(k|k)$.

In contrast to the real-time state estimation problem discussed above, a batch estimation method uses all the m measurements to estimate the states $\mathbf{x}(1)$, $\mathbf{x}(2)$, ..., $\mathbf{x}(L)$. Using a similar notation to denote the estimated state in Table 2.2, batch estimates of the states $\mathbf{x}(1)$, $\mathbf{x}(2)$, ..., $\mathbf{x}(m)$ can be denoted as $\hat{\mathbf{x}}(1|L)$, $\hat{\mathbf{x}}(2|L)$, ..., $\hat{\mathbf{x}}(L|L)$, respectively. One can conclude that the estimation accuracy will be higher in batch estimation compared to that in real-time estimation. The benefit of real-time over batch estimation is that the estimate at time k is instantly available without having to wait for the remaining $L-k$ measurements. For more details about batch estimation (also known as smoothing), the reader is referred to [1].

Table 2.2

Measurements for Recursive Estimation

k	Available Measurements	Real-time Estimate
1	$\mathbf{z}(1)$	$\hat{\mathbf{x}}(1\|1)$
2	$\mathbf{z}(1), \mathbf{z}(2)$	$\hat{\mathbf{x}}(2\|2)$
3	$\mathbf{z}(1), \mathbf{z}(2), \mathbf{z}(3)$	$\hat{\mathbf{x}}(3\|3)$
$\vdots$	$\vdots$	$\vdots$
L	$\mathbf{z}(1), \mathbf{z}(2), \mathbf{z}(3), \ldots, \mathbf{z}(L)$	$\hat{\mathbf{x}}(L\|L)$

The Kalman filter (KF) provides a computationally efficient solution to the real-time estimation problem. Rather than having to process the increasing number of measurements, as listed in Table 2.2, the Kalman filter computes these estimates through a recursive process that requires the same amount of computation regardless of the number of observations. The Kalman filter is derived based on the following assumptions:

1. Linearity: The process model (2.19) and measurement model (2.21) are linear. An approximation solution is available for nonlinear models as described in Section 2.4.

2. Gaussian assumption: The process noise $\mathbf{v}(k)$ and the measurement noise $\mathbf{w}(k)$ are assumed to be zero-mean Gaussian.

3. Known model: The process model (2.19) and measurement model (2.21) are together known as the state-space model (SSM). The parameters of this model are the $m \times m$ state-transition matrix $\mathbf{F}$, the $n \times m$ observation matrix $\mathbf{H}$, the $m \times m$ process-noise covariance matrix $\mathbf{Q}$, and the $n \times n$ measurement-noise covariance matrix $\mathbf{R}$. The KF assumes perfect knowledge of the SSM parameters: $\mathbf{F}$, $\mathbf{H}$, $\mathbf{Q}$, and $\mathbf{R}$.

4. No time correlation: There is no time correlation in the process and measurement noise sequences, that is

$$E\left[\mathbf{v}(i)\mathbf{v}(j)^T\right] = \mathbf{0} \quad \text{when } i \neq j \tag{2.23}$$

$$E\left[\mathbf{w}(i)\mathbf{w}(j)^T\right] = \mathbf{0} \quad \text{when } i \neq j \tag{2.24}$$

where $\mathbf{0}$ is a zero-matrix of appropriate size. In other words, the subsequent process and measurement noise vectors need to be independent.

The KF is summarized in Algorithm 2.1. Here, the state $\mathbf{x}(k)$ is recursively estimated given the measurement $\mathbf{z}(k)$ and the required prior information. The inputs to the KF recursion are the state estimate at time k, $\hat{\mathbf{x}}(k|k)$, corresponding estimation error covariance $\mathbf{P}(k|k)$, and the new measurement $\mathbf{z}(k)$; given these inputs, the KF algorithm produces the updated estimate $\hat{\mathbf{x}}(k+1|k+1)$ and the corresponding estimation error covariance $\mathbf{P}(k+1|k+1)$.

Algorithm 2.1 Kalman Filter
$[\hat{\mathbf{x}}(k+1|k+1), \mathbf{P}(k+1|k+1)] = \text{KF}(\hat{\mathbf{x}}(k|k), \mathbf{P}(k|k), \mathbf{z}(k+1))$

1: *State-prediction:*
$\hat{\mathbf{x}}(k+1|k) = \mathbf{F}\hat{\mathbf{x}}(k|k)$

2: *Covariance of state-prediction error:*
$\mathbf{P}(k+1|k) = \mathbf{F}\mathbf{P}(k|k)\mathbf{F}^T + \mathbf{Q}$

3: *Measurement prediction:*
$\hat{\mathbf{z}}(k+1|k) = \mathbf{H}\hat{\mathbf{x}}(k+1|k)$

4: *Measurement prediction error (innovation/residual):*
$\nu(k+1) = \mathbf{z}(k+1) - \hat{\mathbf{z}}(k+1|k)$

5: *Covariance of the innovation/residual:*
$\mathbf{S}(k+1) = \mathbf{R} + \mathbf{H}\mathbf{P}(k+1|k)\mathbf{H}^T$

6: *Filter gain:*
$\mathbf{W}(k+1) = \mathbf{P}(k+1|k)\mathbf{H}^T\mathbf{S}(k+1)^{-1}$

7: *State-update:*
$\hat{\mathbf{x}}(k+1|k+1) = \hat{\mathbf{x}}(k+1|k) + \mathbf{W}(k+1)\nu(k+1)$

8: *Covariance of the state-update error:*
$\mathbf{P}(k+1|k+1) = \hat{\mathbf{P}}(k+1|k) - \mathbf{W}(k+1)\mathbf{S}(k+1)\mathbf{W}(k+1)^T$

The following MATLAB codes implement the Kalman filtering algorithm summarized in Algorithm 2.1.

Listing 2.1: MATLAB Code for Kalman Filter

```
1  function [xkk, Pkk, NIS] = KF(x0, P0, F, Q, H, R, zk)
2      %state prediction
3      x_pred = F*x0;
4      %state prediction covariance
5      P_pred = F*P0*transpose(F) + Q ;
6      %measurement prediction
7      z_hat = H*x_pred;
8      %innovation
9      inov = zk - z_hat;
10     %innovation covariance
11     S = R + H*P_pred*transpose(H);
12     %kalman gain
13     G = P_pred*transpose(H)*inv(S);
```

```
4      %state update
5      xkk = x_pred + G*inov;
6      %covariance update
7      Pkk = P_pred - G*S*transpose(G);
8      % NIS
9      NIS = transpose(inov)*inv(S)*inov;
0 end
```

Figure 2.3 shows the Kalman filter, summarized in Algorithm 2.1, as a block diagram. The inputs are the previous estimate $\hat{\mathbf{x}}(k|k)$, the estimation error covariance of the previous estimate $\mathbf{P}(k|k)$, and the new measurement $\mathbf{z}(k+1)$. The outputs are the new estimate $\hat{\mathbf{x}}(k+1|k+1)$ and the estimation error covariance of the new estimate $\mathbf{P}(k+1|k+1)$. The Kalman filter recursion continues as the new measurements arrive. It is also important to note that the estimation error covariance is computed independently of the measurements; this indicates the predictability of the uncertainty in the KF estimates.

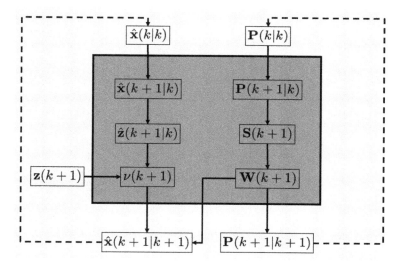

Figure 2.3 One iteration of the Kalman filter. It must be noted that the updated covariance is computed without requiring either the state estimate $\hat{\mathbf{x}}(k|k)$ or the measurement $\mathbf{z}(k+1)$.

Example 2.3 Tracking Resistance Using Kalman Filter – Modeling

The internal resistance of a battery varies due to temperature. The following m measurements were made from a battery at a fixed sampling time of Δ minutes: $z(1), z(2), \ldots, z(L)$. The measured resistances are in milli-ohms (mΩ). The measurements suffer from measurement noise that is zero mean and has a standard deviation of σ_w Ω. Assuming that the standard deviation of the change of resistance variation is σ_v Ω/min^2, implement a Kalman filter to recursively estimate the true resistances of the battery corresponding to each measurement.

We will develop a model that incorporates the fact that the resistance changes over time. Even though the change in resistance is continuous over time, we will develop a discrete model that approximates the changes over a sampling time of Δ minutes. Then we will employ a Kalman filter to recursively estimate the resistance based on the observations.

First, let us define the following state vector

$$\mathbf{x}(k) \triangleq \begin{bmatrix} x(k) \\ \dot{x}(k) \end{bmatrix} \triangleq \begin{bmatrix} R(k) \\ \dot{R}(k) \end{bmatrix} \tag{2.25}$$

where the first element of the state, $R(k)$, denotes the resistance (in milli-ohms or mΩ) at time instant k and the second element, $\dot{R}(k)$, denotes the rate of change of the resistance (in mΩ/min) at time instant k. The following process model is introduced to incorporate the change of resistance over time

$$\mathbf{x}(k+1) = \mathbf{F}\mathbf{x}(k) + \underbrace{\Gamma v(k)}_{\text{process noise}} \tag{2.26}$$

where

$$\mathbf{F} = \begin{bmatrix} 1 & \Delta \\ 0 & 1 \end{bmatrix}, \ \Gamma = \begin{bmatrix} \Delta^2/2 \\ \Delta \end{bmatrix} \tag{2.27}$$

and the process noise $v(k)$ is assumed to be zero-mean Gaussian with variance σ_v^2. The process noise accounts for the uncertainty in the model between time instances k and $k+1$. In this example, the unit of process noise $v(k)$ is in resistance/time^2 (i.e., Ω/min^2). It can be shown [1] that the process noise covariance is

$$\mathbf{Q} = E\left\{\Gamma \mathbf{v}(k)\mathbf{v}(k)^T \Gamma^T\right\} = \begin{bmatrix} \frac{1}{4}\Delta^4 & \frac{1}{2}\Delta^3 \\ \frac{1}{2}\Delta^3 & \Delta^2 \end{bmatrix} \sigma_v^2 \tag{2.28}$$

Let us now write down the observation model. Each measured resistance relates to the state vector as follows:

$$z(k) = \mathbf{H}\mathbf{x}(k) + w(k) \tag{2.29}$$

where

$$\mathbf{H} = \begin{bmatrix} 1 & 0 \end{bmatrix} \tag{2.30}$$

and the measurement noise $w(k)$ (in mΩ) is assumed to be zero-mean Gaussian with variance σ_w^2, that is,

$$\mathbf{R} \triangleq E\left\{w(k)^2\right\} = \sigma_w^2 \tag{2.31}$$

The parameters of the state space model, $\mathbf{F}, \mathbf{Q}, \mathbf{H}$, and $\mathbf{R}$ have been defined in (2.27), (2.28), (2.30), and (2.31), respectively. Given the measurements $z(1), z(2), \ldots, z(L)$, the KF summarized in Algorithm 2.1 can be used to obtain the state estimates $\hat{\mathbf{x}}(1|1)$, $\hat{\mathbf{x}}(2|2), \ldots, \hat{\mathbf{x}}(L|L)$ in a recursive manner. $\square$

In order to start the recursive estimation process using the KF, prior information is needed. Filter initialization is explained as the solution to Example 2.4.

Example 2.4 Tracking Resistance Using Kalman Filter – Implementation

Consider the following data corresponding to Example 2.3. Sampling time is $\Delta = 1$ min., the total number of measurements is $L = 24$, the measurements $z(1), z(2), \ldots, z(24)$, respectively, are given in Table 2.3; the true resistance values $x(k)$ are also given in this table for performance assessment. Implement a Kalman filter to recursively estimate the resistance values $x(k)$ based on the measurements $z(k)$ by assuming the process and measurement noise variances as follows: $\sigma_v^2 = 1$ mΩ/min^2 and $\sigma_w^2 = 4$ mΩ^2.

To answer Example 2.4, the model developed in Example 2.3 will be used. Two filter parameters, the process noise variance σ_v^2 and measurement noise variances σ_v^2, are already given. We will implement a two-step filter initialization procedure [1] to compute the initial estimate $\hat{\mathbf{x}}(2|2)$ and its estimation error covariance $\mathbf{P}(2|2)$. Based on the first two measurements, the estimate of the state $\mathbf{x}(2)$ is

$$\hat{\mathbf{x}}(2|2) = \begin{bmatrix} z(2) \\ \frac{z(2) - z(1)}{\Delta} \end{bmatrix} = \begin{bmatrix} 16.93 \\ 6.6 \end{bmatrix} = \begin{bmatrix} x(2) + w(2) \\ \frac{x(2) - x(1)}{\Delta} + \frac{w(2) - w(1)}{\Delta} \end{bmatrix} \tag{2.32}$$

Table 2.3

True State and Measurements

k	$x(k)$	$z(k)$
1	10.33	6.790
2	16.93	15
3	22.93	21.63
4	28.33	29.23
5	33.13	34.35
6	37.33	35.75
7	40.93	42.58
8	43.93	45.26
9	46.33	46.07
10	48.13	46.22
11	49.33	51.11
12	49.93	49.67
13	49.93	47.90
14	49.33	49.83
15	48.13	49.17
16	46.33	43.40
17	43.93	47.66
18	40.93	36.63
19	37.33	34.05
20	33.13	35.57
21	28.33	28.98
22	22.93	22.07
23	16.93	16.95
24	10.33	11.79

Let us now compute the expectation of $\hat{\mathbf{x}}(2|2)$

$$E(\hat{\mathbf{x}}(2|2)) = \begin{bmatrix} x(2) \\ \frac{x(2)-x(1)}{\Delta} \end{bmatrix} \tag{2.33}$$

where the zero-mean property of the measurement noise (i.e., $E(w(i)) = 0$) is exploited.

The corresponding estimate error is then written as

$$\tilde{\mathbf{x}}(2|2) = \hat{\mathbf{x}}(2|2) - E(\hat{\mathbf{x}}(2|2)) = \begin{bmatrix} w(2) \\ \frac{w(2)-w(1)}{\Delta} \end{bmatrix} \tag{2.34}$$

The initial filter covariance is then

$$\mathbf{P}(2|2) = E\left[\tilde{\mathbf{x}}(2|2)\tilde{\mathbf{x}}(2|2)^T\right] = \begin{bmatrix} \sigma_w^2 & \frac{\sigma_w^2}{\Delta} \\ \frac{\sigma_w^2}{\Delta} & \frac{2\sigma_w^2}{\Delta^2} \end{bmatrix} = \begin{bmatrix} 16 & 16 \\ 16 & 32 \end{bmatrix} \tag{2.35}$$

When the next measurement $z(3)$ arrives, the initial estimate obtained in (2.32) and the corresponding covariance computed in (2.35) are used to call the KF algorithm to compute $\hat{\mathbf{x}}(3|3)$ and $\mathbf{P}(3|3)$ as follows:

$$[\hat{\mathbf{x}}(3|3), \mathbf{P}(3|3)] = \text{KF}(\hat{\mathbf{x}}(2|2), \mathbf{P}(2|2), \mathbf{z}(3)) \tag{2.36}$$

The Kalman filtering iteration continues for the subsequent measurement $z(4)$ arrives, that is,

$$[\hat{\mathbf{x}}(4|4), \mathbf{P}(4|4)] = \text{KF}(\hat{\mathbf{x}}(3|3), \mathbf{P}(3|3), \mathbf{z}(4)) \tag{2.37}$$

and so on.

Remark 2.1 The first two measurements are used for filter initialization. The estimated state using the first two measurements is $\hat{\mathbf{x}}(2|2)$, which is an estimate of $\mathbf{x}(2)$. In a real-time sense, obtaining an estimate of $\mathbf{x}(1)$ is not possible.

Figure 2.4 shows the estimated values of the resistance $\hat{x}(k|k) = \hat{R}(k|k)$ along with the measurements $z(k)$ against time index k. The true value of $x(k)$ from Table 2.3 is also plotted for comparison. The smoothing nature of the KF can be observed from this figure. That is, the Kalman filter estimates are better than the measurements due to the smoothing property found in the filtered estimates. There are other many other advantages of using the Kalman filter; for example, if we were to estimate $x(k)$ and $\dot{x}(k)$ using the two-point approach, the resistance estimates will not be any better than the measurements. Subsequent examples in this chapter will elaborate on the additional benefits of the Kalman filter for recursive state estimation.

Figure 2.5 shows the estimated rate of change of resistance, $\hat{\dot{x}}(k) = \hat{\dot{R}}(k)$ against time index k; it can be observed that it changes from positive to negative around the halfway point reflecting the increase and then decrease of the true resistance. It is also noteworthy that, regardless of the increase/decrease of resistance, the rate of

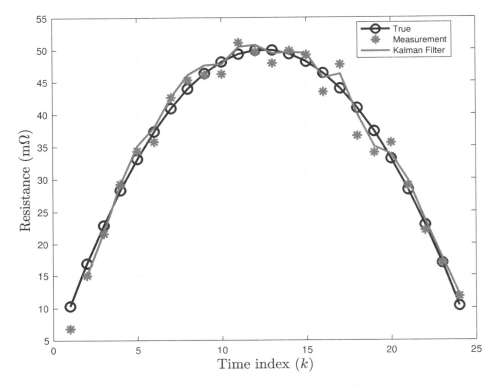

Figure 2.4 Kalman filter estimate of $x(k)$.

change appears to be linear. This offers a glimpse into the benefits of proper modeling. To demonstrate the benefit of this model, a Kalman filter will be modeled without considering the rate of change as a state variable later in Example 2.6.   □

The process and measurement noise variances were provided in Example 2.4. In practice, the measurement noise variance is related to the sensor that is used to obtain the measurements $z(k)$. The measurement noise statistics (mean and variance) could be obtained from the calibration details of the measurement device. Simple experiments can be performed to obtain these as well; for example, by repeatedly measuring a zero-resistance (short circuit), the variance of σ_w^2 can be computed.

Determining the process noise variance is more challenging and indirect compared to that of the measurement noise. Process noise represents the uncertainty in the

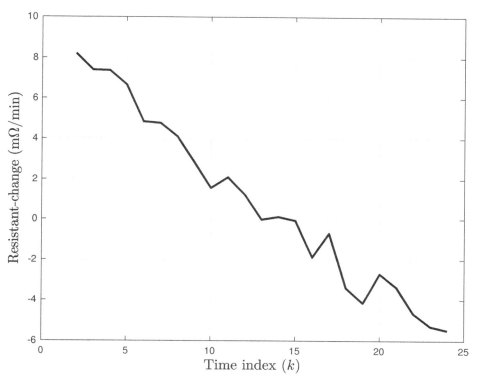

Figure 2.5 Kalman filter estimate of $\dot{x}(k)$.

knowledge of the process model; higher process noise variance implies that there is little knowledge about how the state changes over time. It is obvious that the more knowledge one has about the state transition, the better; in other words, the lower the process noise variance, the better. However, if the selected process noise variance is lower than what is reflected in the data, the Kalman filter runs into a model mismatch problem. This is further illustrated through Example 2.5.

Example 2.5 Kalman Filter Performance Monitoring

Consider the KF implementation in Example 2.4 with the following two cases of model parameters.

- KF-1: The process noise variance is $\sigma_v^2 = 1$.

- KF-2: The process noise variance is $\sigma_v^2 = 0.01$.

Keep all the remaining parameters the same as in Example 2.4. Evaluate the suitability of the process noise assumed by KF-1 and KF-2 based on the normalized innovation squared (NIS).

The NIS is defined as [1]

$$\epsilon_\nu(k) = \nu(k)^T \mathbf{S}(k)^{-1} \nu(k) \tag{2.38}$$

where $\nu(k)$ is the $n_z \times 1$ innovation vector and n_z is the dimension of the observation vector. In the present example, the observations are scalar hence, we have $n_z = 1$.

It can be shown that the NIS, $\epsilon_\nu(k)$, is chi-square-distributed with n_z degrees of freedom. Then we can write with a certain level of confidence α that

$$r_1 \leq \epsilon_\nu(k) \leq r_2 \tag{2.39}$$

where $[r_1, r_2]$ is the two-sided chi-square confidence interval with n_z degrees of freedom. In other words,

$$p(r_1 \leq \epsilon_\nu(k) \leq r_2) = 1 - \alpha \tag{2.40}$$

For $n_z = 1$ degree of freedom, the chi-square-distributed variable $\epsilon_\nu(k)$ will lie between $r_1 = 9.8 \times 10^{-4}$ and $r_2 = 7.4$ with 95% confidence. In other words,

$$p(9.8 \times 10^{-4} \leq \epsilon_\nu(k) \leq 7.4) = 1 - 0.05 = 0.95 \tag{2.41}$$

Figure 2.6 shows the estimates of KF-1 and KF-2. Here, the KF-1 is the same as the one shown in Figure 2.4 in Example 2.4. From the estimates shown in Figure 2.6, it is clear that the process noise assumed by KF-2 is smaller than what it needs to be: smaller process noise results in smoother estimates; however, the estimates are slow to catch up with the actual change in the state.

It is easy to see from Figure 2.6 that the KF-2 implementation has a model parameter mismatch; for trained eyes, it is also easy to see that the process noise variance

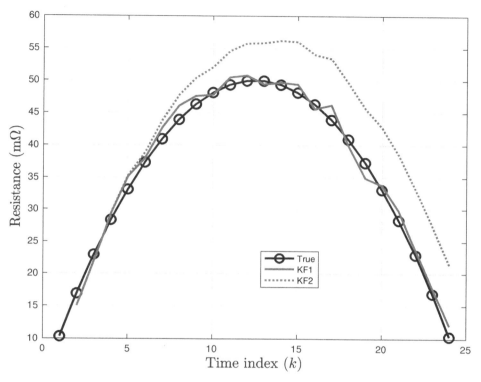

Figure 2.6 Comparison of estimates $\hat{x}(k|k)$ from KF-1 and KF-2.

assumed by the Kalman filter model is lower than what it needs to be. Figure 2.7 compares the NIS of KF-1 and KF-2. The lower and upper limits of the NIS values, $r_1 = 9.8 \times 10^{-4}$ and $r_2 = 7.4$, respectively, are also shown as dashed lines. From this figure, it is easy to see that the NIS corresponding to the KF-2 implementation falls outside the 95% confidence interval most times. It must be emphasized that the NIS values of a filter consistently falling outside the bound indicates a model mismatch; it does not indicate exactly what that model mismatch is. This will bring a significant challenge when it comes to developing a Kalman filtering-based solution to battery state estimation where several model parameters, such as the battery capacity and electrical equivalent circuit model parameters, are not perfectly known.

□

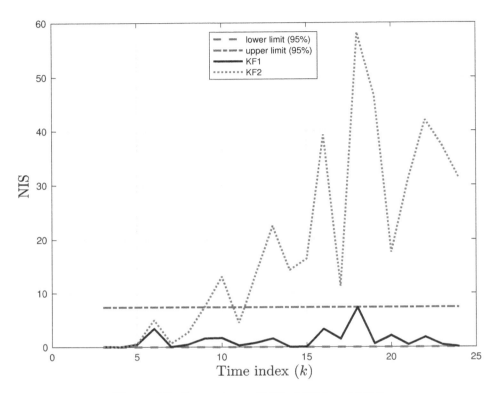

Figure 2.7 Comparison of NIS of KF-1 and KF-2.

Next, we will revisit the state-space modeling from Example 2.3. Here, the selection of a two-state model required some prior understanding of the application in which the Kalman filter is to be used. Instead of selecting a two-state model for the process equation, a one-state model or a three-state model could have been selected. In the one-state model, the state vector would be just the resistance $x(k)$ and in the three-state model, the state vector would consist of three elements, $\mathbf{x}(k) = [x(k) \; \dot{x}(k) \; \ddot{x}(k)]$; here, the third element of the state vector $\ddot{x}(k)$ refers to the rate of change of $\dot{x}(k)$. In kinematic applications, the elements of the three-state vector are the position, velocity, and accelerations; such three-state state vectors are widely applicable in target tracking applications [1].

The NIS can be useful to identify the suitability of the selected state-state model. In Example 2.6, a one-state process equation is employed to estimate the resistances and its performance will be compared to that of the two-state model.

Example 2.6 Tracking Resistance Using Kalman Filter – Alternate Modeling

Consider the problem described in Example 2.3. Develop a Kalman filter to recursively estimate the true resistance values $x(k)$ based on the following process and measurement models:

$$x(k+1) = x(k) + v(k) \qquad (2.42)$$
$$z(k) = x(k) + w(k) \qquad (2.43)$$

where the process noise $v(k)$ is assumed to be zero-mean Gaussian with variance $\sigma_v^2 = 4$ and the measurement noise $w(k)$ is assumed to be zero-mean Gaussian with variance $\sigma_w^2 = 4$.

For this example, the filter parameters can be easily set as

$$\mathbf{F} \to F = 1, \mathbf{H} \to H = 1, \qquad (2.44)$$
$$\mathbf{Q} \to Q = \sigma_v^2, \text{ and, } \mathbf{R} \to R = \sigma_w^2 \qquad (2.45)$$

In this book, the bold symbols are reserved for vectors and matrices. Because the parameters of this model are scalar, regular symbols are used as indicated using the arrows. For filter initialization, just one measurement is enough. It can be shown that

$$\hat{x}(1|1) = z(k), \quad P(1|1) = \sigma_w^2 \qquad (2.46)$$

With $\hat{x}(1|1)$ and $\mathbf{P}(1|1)$ as the prior distribution (i.e., the mean and variance of a Gaussian probability distribution), the Kalman filter in Algorithm 2.1 can be used to compute the state estimates $\hat{\mathbf{x}}(2|2), \ldots, \hat{\mathbf{x}}(24|24)$ based on the measurements $z(2), \ldots, z(24)$ from Table 2.3.

Figure 2.8 shows the Kalman filter estimate along with the true value and measurements for comparison. Based on a visual inspection of the estimates, it can be concluded that the Kalman filter estimates exhibit a bias; it can be further concluded that the sign of the bias changes for the second half of the measurements.

Overall, the filter estimates shown in Figure 2.8 imply that the knowledge of the process model is not accurate enough. This can be further analyzed based on the NIS of

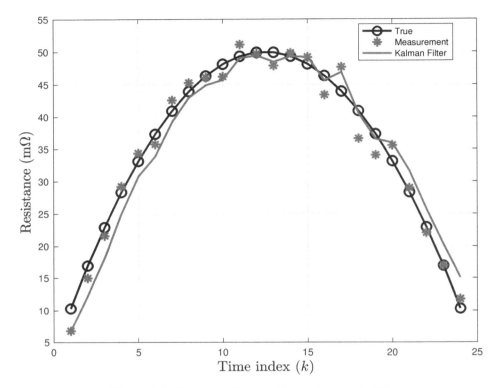

Figure 2.8 One-state Kalman filter estimate of $x(k)$.

the filter, shown in Figure 2.9. The NIS values of the filter are higher than the threshold value at the edges and they are marginally closer to the lower threshold in the middle; indicating a filter that is not consistent. This observation indicates that the state-space modeling needs to be improved.

Table 2.4 compares the mean square error (MSE) of various Kalman filtering approaches discussed in this section through various examples. It can be noticed that, except for the Kalman filter in Example 2.4, the other two filter estimates are worse than the measurements in terms of the MSE. This serves as a reminder that specific details of filter design vary from one application to another. Another important reminder about Kalman filtering applications is that, unlike the examples discussed in this chapter, the true state values are not known. Without the knowledge of the ground truth, it becomes

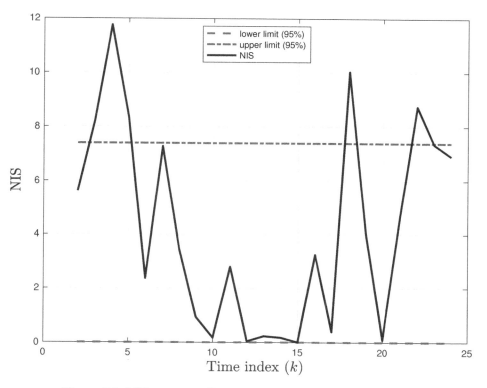

Figure 2.9 NIS corresponding to the one-state KF in Example 2.6.

challenging to assess the performance of the filter. The NIS will be a useful diagnostic measure in practical applications where the true desired state values are never known.

□

2.4   EXTENDED KALMAN FILTER

One of the assumptions of Kalman filtering, which we describe in this section, is that the state-space model needs to be linear. The extended Kalman filter (EKF) provides an approximate solution when the state-space model is nonlinear. A nonlinear state-space

Table 2.4

MSE of Different Estimates

Filter	MSE
Measurements	3.8304
Two state KF in Example 2.4	1.8247
Two state KF-2 in Example 2.5	6.6616
One state KF in Example 2.6	7.1559

model is written in a generic form as follows:

$$\mathbf{x}(k) = f(\mathbf{x}(k-1)) + \mathbf{v}(k) \tag{2.47}$$

$$\mathbf{z}(k) = h(\mathbf{x}(k)) + \mathbf{w}(k) \tag{2.48}$$

which differs from (2.19) and (2.21) due to the fact that the state-transition model and the measurement model are written in a general form. The state-transition model $f(\mathbf{x}(k-1))$ indicates a function of $\mathbf{x}(k-1)$ without the explicit linearity shown by $\mathbf{F}\mathbf{x}(k-1)$ in the earlier version. Similarly, the measurement model replaces the explicitly linear model $\mathbf{H}\mathbf{x}(k)$ with $h(\mathbf{x}(k))$.

Similar to (2.19) and (2.21), the process noise and the measurement noise vectors, in (2.47) and (2.48), respectively, are assumed to be zero-mean Gaussian with covariance matrices $\mathbf{Q}$ and $\mathbf{R}$, respectively. This section summarizes the EKF algorithms and discusses an example to demonstrate its application. For more detailed information on EKF, the readers are directed to [1].

In short, the EKF follows the same KF procedure presented in Algorithm 2.1, after the following linearization steps:

$$\mathbf{F} = \left.\frac{\partial f(\mathbf{x}(k))}{\partial \mathbf{x}(k)}\right|_{\hat{\mathbf{x}}(k|k)}, \mathbf{H} = \left.\frac{\partial h(\mathbf{x}(k))}{\partial \mathbf{x}(k)}\right|_{\hat{\mathbf{x}}(k+1|k)} \tag{2.49}$$

It is possible that the process model is linear and the measurement model is nonlinear or vice versa; in both of these cases, the resulting filtering approach falls under the category of EKF. If the process model is linear and the measurement model is nonlinear, the linearization step is needed only for the measurement model; and, if the measurement model is linear and the process model is nonlinear, the linearization step is needed only for the process model.

2.4.1 Assumptions of the EKF

The EKF algorithm is subject to the following three assumptions:

1. Gaussian assumption: The process noise $\mathbf{v}(k)$ and the measurement noise $\mathbf{w}(k)$ are assumed to be Gaussian.

2. Known model: Similar to the KF, the parameters of this model need to be known, that is, the parameters defining the function $f(\cdot)$, the parameters defining the function $h(\cdot)$, the $m \times m$ process-noise covariance matrix $\mathbf{Q}$, and the $n \times n$ measurement-noise covariance matrix $\mathbf{R}$ are assumed known.

3. No time correlation: Similar to KF, the EKF algorithm is derived under the assumption that the process and measurement noise sequences have no time correlation (i.e., the subsequence process/measurement noise sequences need to be independent).

Next, we will demonstrate the modeling, implementation, and performance monitoring of EKF through the next few examples.

Example 2.7 Nonlinear Filtering Problem

Let us consider the following nonlinear filtering example from [5]

$$x(k) = \frac{x(k-1)}{2} + \frac{25x(k-1)}{1+x(k-1)^2} + 8\cos(1.2k) + v(k) \qquad (2.50)$$

$$z(k) = \frac{x(k)^2}{20} + w(k) \qquad (2.51)$$

where it is assumed that the process and measurement noises are zero-mean i.i.d. Gaussians with variances

$$E\left[v(k)^2\right] = \sigma_v^2 \triangleq Q \quad \text{and} \quad E\left[w(k)^2\right] = \sigma_w^2 \triangleq R \qquad (2.52)$$

Develop an EKF to recursively estimate the states $x(k)$ given the measurements $z(k)$ for $k = 1, 2, \ldots, L$.

For the given example, the nonlinear state-transition equation and the measurement equation are both nonlinear and given by

$$f(x(k-1)) = \frac{x(k-1)}{2} + \frac{25x(k-1)}{1+x(k-1)^2} + 8\cos(1.2k) \qquad (2.53)$$

$$h(x(k)) = \frac{x(k)^2}{20} \qquad (2.54)$$

First, let us perform the linearization step in (2.49) as follows:

$$F_k = \left. \frac{\partial f(x(k-1))}{\partial x(k-1)} \right|_{\mathbf{x}(k)=\hat{x}(k-1|k-1)} \qquad (2.55)$$

$$F_k = \frac{1}{2} + \frac{25\left(1 - \hat{x}(k-1|k-1)^2\right)}{\left(1 + \hat{x}(k-1|k-1)^2\right)^2} \qquad (2.56)$$

where F_k is the linearized state transition model at time k. Linearization for the measurement model is given as

$$H_k = \left. \frac{\partial h(x(k))}{\partial x(k)} \right|_{\mathbf{x}(k)=\hat{x}(k|k-1)} = \frac{\hat{x}(k|k-1)}{10} \qquad (2.57)$$

where H_k is the linearized observation model.

Now, we will follow the KF procedure presented in Algorithm 2.2 to rewrite the EKF procedure for the nonlinear state-space model given in (2.47) and (2.48); Algorithm 2.2 summarizes the EKF algorithm corresponding to Example 2.7. It must be emphasized that, except for the linearization step, the EKF procedure is identical to that of the Kalman filter.

The following MATLAB codes implement the Kalman filtering algorithm summarized in Algorithm 2.2.

Listing 2.2: MATLAB Code for Extended Kalman Filter

```
1 function [xkk, Pkk, NIS] = EKF(xkk, Pkk, Q, R, zk, k)
2     % state prediction
3     x_pred = xkk/2 + 25*xkk/(1+power(xkk,2)) + 8*cos(1.2*k);
4     % Linearization to obtain F:
5     F = 25/(power(xkk,2) + 1) - ...
6         (50*power(xkk,2))/power((power(xkk,2) + 1),2) + 1/2;
7     % Covariance of state-prediction error
8     P_pred = F*Pkk*transpose(F) + Q;
9     % Measurement prediction
```

Algorithm 2.2 Extended Kalman Filter

$[\hat{x}(k+1|k+1), P(k+1|k+1)] = \text{EKF}(\hat{x}(k|k), P(k|k), z(k+1))$

1: *State-prediction:*
$$\hat{x}(k+1|k) = \frac{\hat{x}(k|k)}{2} + \frac{25\hat{x}(k|k)}{1+\hat{x}(k|k)^2} + 8\cos(1.2k)$$

2: *Linearization to obtain F:*
$$F_k = \frac{1}{2} + \frac{25\left(1-x(k-1)^2\right)}{(1+x(k-1)^2)^2}$$

3: *Covariance of state-prediction error:*
$$P(k+1|k) = F_k P(k|k) F_k^T + Q$$

4: *Measurement prediction:*
$$\hat{z}(k+1|k) = \frac{x(k)^2}{20}$$

5: *Measurement prediction error (innovation/residual):*
$$\nu(k+1) = z(k+1) - \hat{z}(k+1|k)$$

6: *Linearization to obtain H:*
$$H_k = \hat{x}(k+1|k)/10$$

7: *Covariance of the innovation/residual:*
$$S(k+1) = R + H_k P(k+1|k) H^T$$

8: *Filter gain:*
$$W(k+1) = P(k+1|k)H^T S(k+1)^{-1}$$

9: *State-update:*
$$\hat{x}(k+1|k+1) = \hat{x}(k+1|k) + W(k+1)\nu(k+1)$$

10: *Covariance of the state-update error:*
$$P(k+1|k+1) = P(k+1|k) - W(k+1)S(k+1)W(k+1)^T$$

```
z_hat = (power(x_pred,2))/(20);
% Measurement prediction error (innovation/residual)
inov = zk - z_hat;
% Linearization to obtain H
H = x_pred/10;
% Covariance of the innovation/residual
S = R + H*P_pred*transpose(H);
% Filter gain
G = (P_pred*transpose(H))/S;
% State-update
xkk = x_pred + G*inov ;
% Covariance of the state-update error
Pkk = (1-G*H)*P_pred*(transpose(1- G*H)) + G*R*transpose(G);
% Normalized Innovation Squares (NIS)
NIS = (power(inov,2))/S;
end
```

☐

The next example illustrates further EKF implementation using data.

Example 2.8 Implementing the EKF

Consider the nonlinear SSM provided in Example 2.7. Table 2.5 provides data generated according to the model in (2.50) and (2.51) where the process and measurement noise variances are assumed to be $\sigma_v^2 = 0.1$ and $\sigma_r^2 = 0.5$, respectively. Use EKF in Algorithm 2.2 to compute $\hat{x}(1|1), \ldots, \hat{x}(20|20)$ based on the given measurements $z(1), \ldots, z(20)$.

Table 2.5

True State and Measurements

k	$x(k)$	$z(k)$
1	2.899	0.623
2	3.457	-0.434
3	1.043	-1.240
4	13.568	9.876
5	16.525	15.272
6	14.689	9.264
7	5.419	2.375
8	-0.891	0.197
9	-14.17	10.312
10	-1.870	0.095
11	-4.897	1.261
12	-9.677	5.689
13	-15.346	12.261
14	-13.251	8.019
15	-3.248	0.349
16	-0.763	0.458
17	-12.106	6.306
18	-15.806	12.634
19	-15.111	12.329
20	-5.385	1.039

There are $m = 20$ measurements given. The goal here is to find the estimates $\hat{x}(1|1)$, $\hat{x}(2|2)$, ..., $\hat{x}(20|20)$ and the associated estimation error covariances $P(1|1)$, $P(2|2)$, ..., $P(20|20)$.

The first step is to identify parameters of the SSM F, H, Q, and R. Because both the process model (2.50) and the measurement model (2.50) are nonlinear, F_k and H_k will be computed according to (2.56) and (2.57), respectively. The other two parameters are $Q = \sigma_v^2 = 0.1$ and $R = \sigma_r^2 = 0.5$ as provided; in this example, the process noise variance Q and measurement noise variance R are time-independent.

The next step is filter initialization. Because the state vector has only one element, the EKF can be initialized using just one measurement (i.e., we will use $z(1)$ to initialize the filter). Let us denote the initial state and its uncertainty (covariance) as $\hat{x}(1|1)$ and $P(1|1)$, respectively.

Remark 2.2 The prior information is referred to as distribution because, in Kalman filtering, one is estimating the posterior probability distribution of the state of interest. Under the assumptions of the Kalman filter (and the extended Kalman filter), the estimated state and the prior information are Gaussian distributions. A Gaussian distribution can be defined by just two parameters: the mean and the covariance. The prior information denotes a Gaussian distribution with mean $\hat{x}(1|1)$ and (co)variance $P(1|1)$. Each subsequent estimates of the (extended) Kalman filter are also Gaussian distribution with mean $\hat{x}(k|k)$ and (co)variance $P(k|k)$ for $k = 2, \dots, m$.

We will employ a maximum likelihood (ML) estimation approach for filter initialization. Based on the observation model (2.51), the likelihood function $\Lambda(x)$ is given as

$$\Lambda(x) = \frac{1}{\sqrt{2\pi}\sigma_r} e^{-\frac{(z-x^2/2)^2}{2\sigma_r^2}} \tag{2.58}$$

where $x(k)$ and $z(k)$ are written as x and z, respectively. The time index k is omitted for convenience during the derivation procedure detailed next. The ML estimate is obtained by setting the derivative of the cost function to zero

$$\frac{\partial \ln \Lambda(x)}{\partial x} = 0 \Rightarrow (z - x^2/2)\, x/\sigma_r^2 = 0 \tag{2.59}$$

Assuming that $x \neq 0$, the ML estimate of x is then

$$\hat{x}_{\mathrm{ML}} = \pm\sqrt{20z} \tag{2.60}$$

For the given problem, there is uncertainty about the sign. One may need to use other available information to select a suitable sign for the estimate. One of the advantages of Kalman filtering is that incorrect initialization has little effect on the filter output in the long run. This will be further illustrated in subsequent examples. The variance of the ML estimate is defined as

$$\sigma_{\mathrm{ML}}^2 = E\left((\hat{x}_{\mathrm{ML}} - E(\hat{x}_{\mathrm{ML}}))^2\right) \tag{2.61}$$

where $E(\cdot)$ denotes expectation operator. It may not be feasible to obtain a closed-form expression for the variance of the estimate. Instead, we will derive the theoretical lower bound on the variance, also known as the Cramer-Rao lower bound (CRLB), and use it to initialize the filter. From (2.58), the log-likelihood function is

$$\ln \Lambda(x) = \left(\frac{1}{\sqrt{2\pi}\sigma_w}\right) - \frac{(z - x^2/2)^2}{2\sigma_w^2} \tag{2.62}$$

The first and second derivates of the log-likelihood function are

$$\frac{\partial \ln \Lambda(x)}{\partial x} = \frac{(z - x^2/2)x}{\sigma_w^2} \tag{2.63}$$

$$\frac{\partial^2 \ln \Lambda(x)}{\partial^2 x} = \frac{(z - 3x^2/2)}{\sigma_w^2} \tag{2.64}$$

The Fisher information is

$$J = -E\left(\frac{\partial^2 \ln \Lambda(x)}{\partial^2 x}\right) = -E\left(\frac{(z - 3x^2/20)}{10\sigma_w^2}\right) \tag{2.65}$$

$$J = -E\left(\frac{(x^2/20 + w - 3x^2/20)}{10\sigma_w^2}\right) = \frac{x^2}{100\sigma_w^2} \tag{2.66}$$

The CRLB is then described as follows

$$E((x - \hat{x})^2) \geq \mathrm{CRLB} = J^{-1} = \frac{100\sigma_w^2}{x^2} \tag{2.67}$$

Based on the discussion so far, the initial estimate of x can be obtained based on (2.60) from the first measurement as follows

$$\hat{x}(1|1) = \sqrt{20z(1)} = \sqrt{20 * 0.623} = 3.5299 \tag{2.68}$$

The initial filter covariance is assigned based on (2.67) as follows

$$\hat{P}(1|1) = \frac{\sigma_w^2}{\hat{x}(1|1)^2} = \frac{0.25}{3.5299^2} = 0.0201 \tag{2.69}$$

With $\hat{x}(1|1)$ and $P(1|1)$ as the initial estimate and its variance, respectively, the EKF algorithm (see Algorithm 2.2) is called for the remaining data for $k = 2, 3, \ldots, 20$. Figure 2.10 shows the EKF estimates $\hat{x}(k|k)$ along with the true values of $x(k)$ and the measurements $z(k)$. Most of the NIS values shown in Figure 2.11 are within the 95% confidence limit, indicating that the filter is properly implemented.

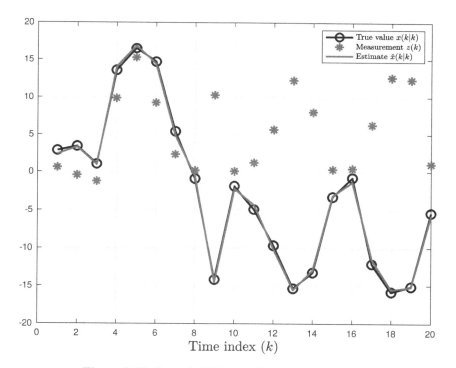

Figure 2.10 Extended Kalman filter estimate of $x(k)$.

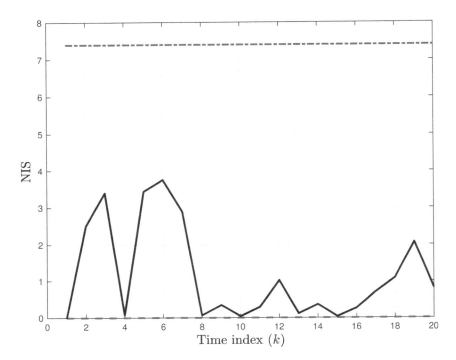

Figure 2.11 Extended Kalman filter estimate of $x(k)$.

Example 2.9 Effect of Wrong Filter Initialization

Consider the nonlinear SSM discussed in Example 2.8. Instead of estimating the prior distribution from the measurement $z(1)$ for filter initialization, initialize the filter using two arbitrary distributions as given below and compare the performance of the resulting estimates.

- EKF-1: $\hat{x}(1|1) = 1, P(1|1) = 10$.

- EKF-2: $\hat{x}(1|1) = 100, P(1|1) = 10$.

From Table 2.5, it can be verified that the true initial state is $\hat{x}(1) = 2.899$. The initial estimates assumed by both EKF-1 and EKF-2 are away from the true initial value;

however, the EKF-2 initial estimate is significantly far from the true value. The selection of initial covariance of $P(1|1) = 10$ is much higher than the practical value obtained in (2.69). The selection of higher initial covariance of $P(1|1) = 10$ informs the filter that there is higher uncertainty in the selected initial estimate $\hat{x}(1)$. Figure 2.12 shows the estimates of EKF-1 and EKF-2. It can be seen that both EKF-1 and EKF-2 estimates converge to the true value within a few iterations. This indicates the insignificance of wrong filter initialization in the long run.

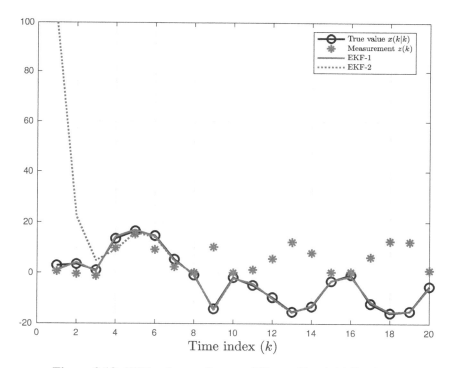

Figure 2.12 EKF estimates for two different filter initializations.

$\square$

The final example of this chapter gives more insights into filter performance monitoring.

Example 2.10 The Effect of Model Mismatch in EKF

Consider the nonlinear SSM discussed in Example 2.8. Implement an EKF by considering the following observation model:

$$x(k) = x(k-1) + \frac{25x(k-1)}{1 + x(k-1)^2} + 8\cos(1.3k) + v(k) \qquad (2.70)$$

Assume that the process and measurement noise variances are the same, that is, $\sigma_v^2 = 0.1$ and $\sigma_r^2 = 0.5$, respectively, and that the measurement model is the same as the one given in Example 2.8. Use EKF in Algorithm 2.2 to compute $\hat{x}(1|1), \ldots, \hat{x}(20|20)$ based on the given measurements $z(1), \ldots, z(20)$ provided in Table 2.5.

Here, one can notice that the only change in the process model is $8\cos(1.2k) \rightarrow 8\cos(1.3k)$. Only the first step needs to be changed in the EKF Algorithm 2.2. All the remaining steps of the EKF are the same.

Figure 2.13 shows the EKF estimates based on the same measurements provided in Table 2.5. Here, filter performance can be seen to deviate from the true value. If it is not for the true value in Figure 2.13, it would have been impossible to know that the filter outputs are incorrect. The corresponding NIS values shown in Figure 2.14 far exceed the 95% limits; this indicates that there is a mismatch between the assumed parameters at the filter and the reality based on the observed data.

□

Example 2.10 serves as a reminder that, in practical applications, the true value of the state of interest is not known. The performance of the filter needs to be continuously monitored based on metrics, such as the NIS. When designing filtering solutions for new applications, the suitability of the estimated model parameters can be tested in a simulated environment. Also, it is important to emphasize that all model mismatches affect the filter performance the same. Some model mismatches affect the filter performance too severely than others. Question 2.2 in Section 2.7 illustrates an example where the model mismatch does not significantly affect the filter performance.

2.5 CONCLUSIONS

This chapter summarized some required concepts in estimation and filtering. Particularly, the least squares estimator and the Kalman filter are described with the help of examples. This chapter introduced how to develop a Kalman filtering solution to a given

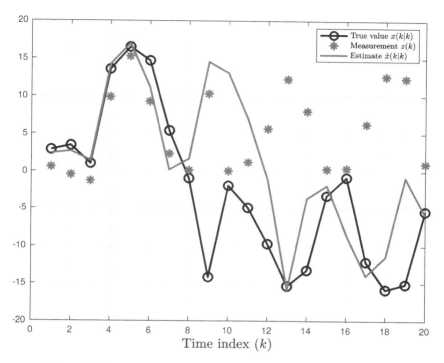

Figure 2.13 EKF estimates in the presence of a model mismatch.

problem. Implementation of a Kalman filter was demonstrated using an example and data. Some relevant MATLAB codes were provided to implement the examples provided in this chapter. The difference between the Kalman filter and the extended Kalman filter was demonstrated through various examples. Finally, this chapter emphasized the importance of continuous monitoring of filter performance.

2.6   BIBLIOGRAPHICAL NOTES

This chapter assumed undergraduate-level knowledge in matrix linear algebra and probability theory. More detailed derivations and analyses of the estimation and filtering concepts can be found in [1]. In addition to estimation theory, the remaining chapters of this book will utilize concepts of linear algebra, probability theory, and numerical

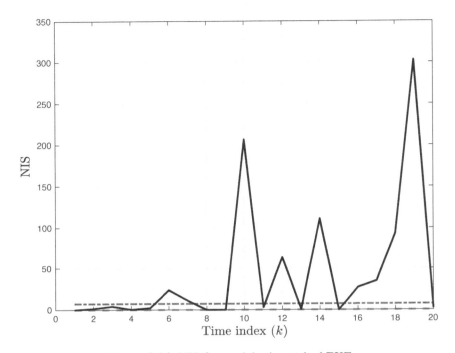

Figure 2.14 NIS for model mismatched EKF.

methods; the reader is referred to [2], [3], and [6] to review concepts in linear algebra, probability theory, and numerical methods, respectively.

2.7 PROBLEMS

Question 2.1 Parameter Estimation and State Estimation

In Example 2.2, it is assumed that all the m measurements were taken fast enough that neither the EMF nor the resistance changed during the course of taking those m measurements. Let us denote that these batches of m measurements were taken at every time instant $k = 1, \ldots, L$. Model the change of EMF and the resistance from one time instant to another using a process model and develop a KF track EMF and the resistance over time.

Question 2.2 More on Model Mismatch

Not all model mismatches equally affect the filter performance. Implement the EKF in Example 2.7 by assuming the following process model and measurement models using the data provided in Table 2.5.

$$x(k) = x(k-1) + \frac{20x(k-1)}{1 + x(k-1)^2} + 8\cos(1.2k) + v(k) \tag{2.71}$$

The mismatch here is that the coefficient changed from 25 to 20. Keep all other parameters of the model the same as in Examples 2.7 and 2.8.

References

[1] Y. Bar-Shalom, X.R. Li, and T. Kirubarajan, *Estimation with Applications to Tracking and Navigation: Theory, Algorithms and Software*, John Wiley & Sons, New York, 2004.

[2] G. Strang, *Introduction to Linear Algebra*, 5th ed., Cambridge University Press, Cambridge, UK, 2016.

[3] J. A. Gubner, *Probability and Random Processes for Electrical and Computer Engineers,* Cambridge University Press, Cambridge, UK, 2006.

[4] K. B. Petersen, and M. S. Pedersen, *The Matrix Cookbook*, Technical University of Denmark, 2008.

[5] M. S. Arulampalam, S. Maskell, N. Gordon, and T. Clapp, "A tutorial on particle filters for online nonlinear/non-Gaussian Bayesian tracking," *IEEE Transactions on Signal Processing* Vol. 50, No. 2, pp. 174-188, 2002.

[6] S. C. Chapra and R. P. Canale, *Numerical Methods for Engineers*, McGraw-Hill, New York, 2011.

Chapter 3

Battery Modeling

3.1 INTRODUCTION

Chapter 1 discussed the need for a BMS, particularly when it comes to Li-ion batteries. Battery modeling is a necessary step in BMS development in which we try to represent the functioning of a battery through mathematical models. Two important functions of a battery are charging and discharging. In addition to that, a battery goes through other functionalities, such as aging. A good battery model will represent all these different functionalities that happen within a battery cell.

Before we get into battery modeling, let us take a brief look at the internal structure and workings of a Li-ion battery. Figure 3.1 shows the four main components of a Li-ion rechargeable battery: anode, cathode, separator, and electrolyte.

The negative electrode (anode) delivers electrons to the circuit during discharge and it receives electrons during charge. In most Li-ion cells, the anode is composed of graphite, a hexagonal structure made of 6 carbon atoms that are tightly bonded together to form one layer called graphene. Graphene layers are loosely held to each other through weak Van der Waals forces. The Li-ions intercalate between the graphene layers.

Current collectors take no part in the chemical reaction. They are responsible for delivering or receiving electrons from the circuit. Their sole purpose is to help collect electrons after the reaction happens. As a result, it can help reduce the internal resistance of the battery. In most Li-ion cells, negative current collectors are made of copper. The positive current collector is usually made of aluminum.

The positive electrode (cathode) is responsible for receiving electrons from the circuit during discharge and delivering them during charge. There are multiple chemistries that can be used in the cathode as active material in Li-ion cells, such as lithium cobalt oxide, lithium nickel manganese cobalt oxide, lithium manganese oxide, lithium nickel

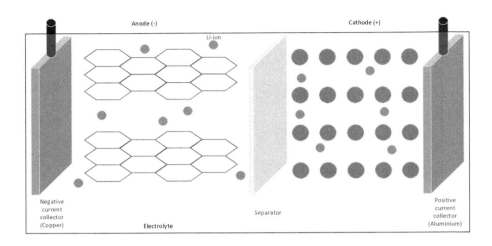

Figure 3.1 Rechargeable lithium-ion battery cell.

cobalt aluminum oxide, and lithium iron phosphate. The principle of operation is the same for all these chemistries where the material is made of a layer-like structure that stays the same through the discharge and charge process while Li-ions can intercalate between this layered structure.

A separator is used to prevent any direct contact between the anode and the cathode that would lead to a short circuit inside the cell. The separator should also be an electrical insulator so that electrons are prevented from traveling internally between the anode and cathode without going through the external circuit. It should also allow for the flow of Li-ions that use electrolytes as their medium. For this, the separator is made of pores to create a passage of the electrolytes. Also, the pores need to be large enough to provide sufficient volume for an electrolyte for high-rate conduction but small enough to prevent dendrites.

Finally, the electrolyte is responsible for the flow of Li-ions through the cell between the anode and cathode. It is comprised of a solvent and salt; the most common salt used is lithium hexafluorophosphate, which is dissolved in the solvent such as ethylene carbonate or diethyl carbonate.

The driving force of charging and discharging activities is the difference in electrochemical potentials between cathode and anode materials and the change in their oxidation state. During discharging, the reduced form of graphite/lithium compound loses an electron; this results in a Li-ion being freed up. The released electron reaches the cathode side where the oxidized form of cathode material (for example, $LiCoO_2$) receives the electron and turns into a more reduced form; it is then compensated by the charge with the Li-ion that travels from the anode. This process is reversed during charging.

The charging and discharging mechanism of a rechargeable battery is affected by external factors, such as temperature, usage patterns, and age. For example, at low temperatures the mobility of electrons and lithium ions is restricted, resulting in high electrical resistance; this resistance is often referred to as charge transfer resistance. When the battery is repeatedly charged and discharged (i.e., cycled), it forms a solid electrolyte interface (SEI); this resistance is known as SEI resistance. The depletion of lithium ions over time results in the reduction of battery capacity. In order to capture the change in battery response over time, it is important to accurately model the battery. The remainder of this chapter provides the details of electrical equivalent circuit models.

A battery model attempts to emulate the behavior of a battery as close to reality as possible. The model allows computing directly unmeasurable states of the battery, such as SOC, SOH, TTS, and RUL (see Figure 3.2). Based on these computed states, the BMS issues electronic control signals to maintain the state of the battery at favorable levels. Present-day battery models are very reliable to predict instantaneous states of the battery, such as SOC, state of power, and TTS. Models to predict the accumulative states of the battery, such as power fade, capacity fade, SOH, and RUL, are still in the early stages. The remainder of this chapter presents well-known models for battery modeling.

3.2 ELEMENTS OF ELECTRICAL EQUIVALENT CIRCUIT MODELS

The requirement in battery modeling, with respect to battery management, is to have a model that allows observing important states and parameters pertaining to the state of charge and state of health of the battery based on noninvasive measurements. Possible noninvasive measurements are the voltage across the battery terminals, the current through the battery, and the temperature on the battery surface. Hence, electrical equivalent circuit models are widely adopted to monitor the battery. It is also possible to noninvasively measure humidity and physical expansion of battery cells. It is possible to model these extra measurements to observe various useful states related to the safety and health of a batter pack. However, the discussion in this chapter and this book, in general, is limited to electrical equivalent circuit models (ECMs).

In the remainder of this section, two different equivalent circuit models are presented and discussed: the direct current (DC) equivalent circuit model and the alternating current (AC) equivalent circuit model.

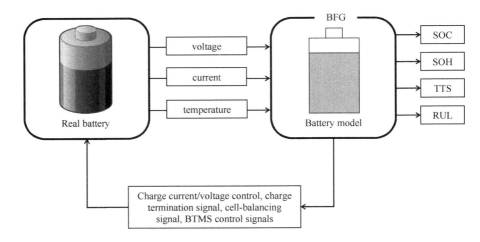

Figure 3.2 Role of battery modeling in battery management systems.

3.2.1   DC Equivalent Circuit Model

Figure 3.3 shows the DC equivalent circuit model (DC-ECM) of a battery. The electromotive force of the battery is denoted by EMF, $h(k)$ denotes hysteresis, R_0, R_1, R_2 denote ohmic resistance, and C_1, C_2 denote capacitance. The voltage across the battery terminals and the current through the battery is denoted by $v(k)$ and $i(k)$, respectively. The current through resistances R_1 and R_2 are denoted by $i_1(k)$ and $i_2(k)$, respectively. This makes the current through capacitances C_1 and C_2 to be $i(k) - i_1(k)$ and $i(k) - i_2(k)$, respectively. Here, the time is denoted in discrete domain (i.e., $i(k)$ denotes the current measured at time k). The discussions in this chapter also assume uniform sampling (i.e., the time between the sample k and sample $k - 1$ is assumed to be the same for all values of k).

Based on the notations described so far, and shown in Figure 3.3, the measured voltage across the battery terminals can be written as

$$v(k) = \text{EMF} + h(k) + i(k)R_0 + i_1(k)R_1 + i_2(k)R_2 \qquad (3.1)$$

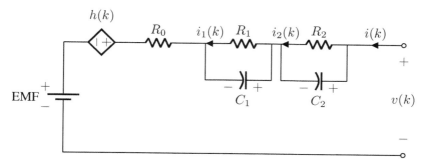

Figure 3.3 DC equivalent circuit model.

The currents $i_1(k)$ and $i_2(k)$ can be derived in terms of $i(k)$ as follows

$$i_1(k+1) = \alpha_1 i_1(k) + (1 - \alpha_1)i(k) \tag{3.2}$$
$$i_2(k+1) = \alpha_2 i_2(k) + (1 - \alpha_2)i(k) \tag{3.3}$$

where $\alpha_1 = e^{-\frac{\Delta}{R_1 C_1}}$, $\alpha_2 = e^{-\frac{\Delta}{R_2 C_2}}$ and Δ denotes the sampling time.

The quantities R_0, R_1, R_2, C_1, and C_2 in Figure 3.3 are shown without a time index (k), indicating that these are the ECM parameters. Hysteresis is shown with time index (k), indicating that hysteresis is not a constant parameter; rather, it is a quantity that changes with time. Later, the hysteresis will $h(k)$ will be introduced as a function of the present and past values of current, the state of charge of the battery, and hysteresis. It must be noted that the ECM parameters may change with the temperature and state of charge of the battery. The battery impedance significantly increases at low temperatures and low states of charge regions. The discussion about the ECM in this chapter assumes that the temperature is constant.

3.2.2 AC Equivalent Circuit Model

Figure 3.4 shows the AC equivalent circuit model of the battery that is also known in the literature as the adaptive Randles equivalent circuit model (AR-ECM). In addition to the EMF component, the AR-ECM consists of the following elements: stray inductance L, ohmic resistance R_Ω, solid electrolytic interface (SEI) resistance R_{SEI}, solid electrolytic interface (SEI) capacitance C_{SEI}, double layer (DL) resistance R_{DL}, double layer (DL) capacitance C_{DL}, and Warburg impedance Z_w. In this chapter, the units of resistance, inductance, and capacitances are referred to in ohms (Ω), henrys (H), and farads (F) unless otherwise explicitly mentioned.

The AC and DC equivalent models are very similar as they both represent the same battery. One significant exception in the AC equivalent circuit model is the Warburg impedance, denoted as Z_w in Figure 3.4. The Warburg impedance is written as

$$Z_w \triangleq Z_w(j\omega) = \sigma \sqrt{\frac{2}{j\omega}} \tag{3.4}$$

where ω indicates angular frequency. It can be noticed that the Warburg impedance is significant only at very low frequencies. At high frequencies (i.e., when ω is significantly high), the effect of Warburg impedance becomes insignificant and the AC equivalent circuit starts to look very similar to its DC counterpart.

Remark 3.1 The expression for Warburg impedance in (3.4) is only an approximation. For a detailed discussion about alternate, more accurate forms, the reader is referred to [1].

It is also possible to add the hysteresis component to the AC equivalent model. Similarly, the inductive component can be added to the DC equivalent circuit models as well. However, the effect of inductance is only observed in very high frequencies limiting it to laboratory studies that most often employ frequency-domain analysis in batteries.

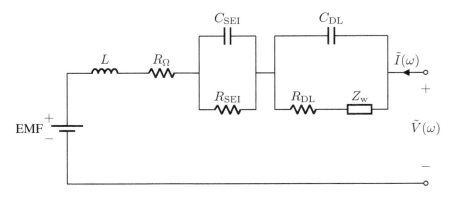

Figure 3.4 AC equivalent circuit model.

Let us denote the voltage across the battery and the current through it as

$$V(\omega) = V_{dc} + \tilde{V}(\omega) \tag{3.5}$$

$$I(\omega) = I_{dc} + \tilde{I}(\omega) \tag{3.6}$$

where $V(\omega)$ and $I(\omega)$ denote the voltage and current measurements consisting of the DC component (V_{dc}, I_{dc}) and AC component $(\tilde{V}(\omega), \tilde{I}(\omega))$. The AC impedance of the battery can be written as

$$
Z(j\omega) = \frac{\tilde{V}(\omega)}{\tilde{I}(\omega)} = j\omega L + R_\Omega + \cfrac{1}{\frac{1}{R_{SEI}} + j\omega C_{SEI}} + \cfrac{1}{\frac{1}{R_{CT}+Z_w(j\omega)} + j\omega C_{DL}}
$$
$$
= j\omega L + R_\Omega + \frac{R_{SEI}}{1 + j\omega R_{SEI}C_{SEI}} + \frac{R_{CT} + Z_w(j\omega)}{1 + j\omega\left(R_{CT} + Z_w(j\omega)\right)C_{DL}} \quad (3.7)
$$

By measuring the AC impedance across a battery, the AR-ECM model parameters can be estimated. Such frequency-domain impedance analysis is known as electrochemical impedance spectroscopy (EIS). In EIS, the Nyquist plot is made by plotting the real part of $Z(j\omega)$ against the imaginary part of it. All the parameters of the AC equivalent circuit model shown in Figure 3.4 can be recovered from a Nyquist plot. The following MATLAB codes generate the Nyquist plot for the ECM parameters corresponding to Figure 3.4.

Example 3.1

The parameters of the adaptive Randles ECM shown in Figure 3.4 are: $L = 4 \times 10^{-6}$ H, $R_\Omega = 0.5\Omega$, $R_{SEI} = 0.1\Omega$, $C_{SEI} = 0.2$F, $R_{CT} = 0.5\Omega$, $C_{DL} = 100$ F, and $\sigma = 0.005$. Generate the Nyquist plot corresponding to this battery.

The following MATLAB codes will generate the Nyquist plot shown in Figure 3.5 based on the AC ECM model parameters provided in Example 3.1.

Listing 3.1: **Nyquist Plot**

```
clear all; clc; close all
% file name: 'NyquistPlot.m'
% initialize AC-ECM model parameters
R_Omega = 0.5; R_SEI = 0.1;  C_SEI = 0.2;
R_CT = .5; C_DL = 100; L= 4e-6; Sig = 0.005;
% create frequency array
fmin = .00001; fmax = 1000; frange = [];
for i=1:ceil(log10(fmax/fmin))
    fr = fmin:fmin/4:10*fmin;
    frange = [frange fr];
    fmin = 10*fmin;
end
w = 2*pi*frange; % angular frequency
```

```
14 Zw = Sig*sqrt(2./(j*w)); % Warburg Imp.
15 Z = j*w*L + R_Omega   + ...
16    1./(1./(R_CT+Zw)+j*w*C_DL)+ ...
17    1./(1/R_SEI+j*w*C_SEI);
18 plot(real(Z), -imag(Z))% Nyquist plot
19
20 h = figure; hold on; grid on; box on;
21 plot(real(Z), -imag(Z), 'linewidth', 3, 'Color', [0 0 1])
22 xlabel('Re($Z_w(\omega)$)', 'Interpreter', 'Latex', 'fontsize', 15)
23 ylabel('Im($Z_w(\omega)$)', 'Interpreter', 'Latex', 'fontsize', 15)
24 xlim([R_Omega-.1 max(real(Z))+.1])
25 filename = '../../_figures/NyquistPlot'
26 print(h, '-depsc', filename)
```

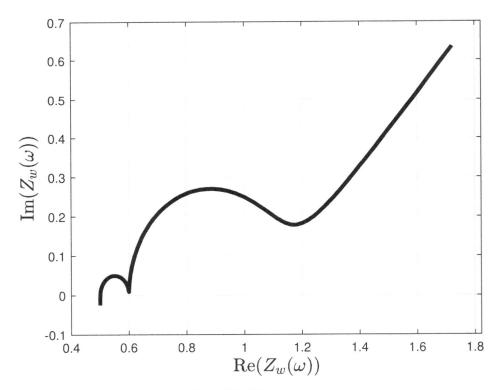

Figure 3.5 Nyquist plot.

□

Remark 3.2 It must be noted that both the DC and AC equivalent circuit models introduced so far did not represent one important parameter of a battery: the battery capacity. In the next section, the role of battery capacity within the battery ECM will be made clear. Additional discussions on battery capacity are provided in Section 3.5.

3.3 REDUCED-ORDER MODELS

As it is demonstrated through a simulation example in Section 3.2.2, the AC equivalent model requires the application of wide-ranging frequency to obtain the Nyquist spectrum. Consequently, the applications of AC equivalent circuit models are limited to laboratory analysis of batteries. Battery management systems need to operate in real-time with the help of opportunistic measurements of voltage and current. Practical battery management systems often use reduced-order models made of selected elements of the DC equivalent circuit model to represent the behavior of the battery. This section provides further details about each component of the DC equivalent circuit model.

3.3.1 Ideal Battery Model

An ideal battery would have the following features:

1. Constant voltage source: It indefinitely acts as a constant voltage source regardless of the amount of Coulombs taken from it.
2. Zero internal resistance: The voltage across the battery terminals remains the same regardless of the magnitude of the current.
3. Zero hysteresis: There is no hysteresis (memory) effect in the voltage and current measurements.

Figure 3.6 shows the equivalent circuit model of an ideal battery. The ideal battery model is used in circuit analysis courses; however, it has little use in the study of battery management systems.

In contrast to the ideal battery model shown in Figure 3.6, real-world battery cells exhibit the following features:

1. Limited voltage source: In practice, the battery voltage does not stay at a constant $v = EMF$ until it empties. As the charge measured in Coulombs is taken away from the battery the EMF voltage gradually drops. The EMF voltage is widely known as the open-circuit voltage (OCV) and is denoted by V_o in this book. The OCV is found to have a monotonous relationship with the state of charge of the battery. As the battery is discharged, its state of charge decreases and so does the OCV.

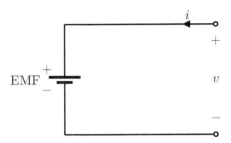

Figure 3.6 Ideal battery model. An ideal battery is one that functions indefinitely as a constant voltage source; that is, the measured terminal voltage would always be $v = \text{EMF}$ regardless of the current i.

2. Electrical resistance effect: In addition to the state of charge-dependent OCV, the measured terminal voltage is observed to be affected by voltage drop due to electrical resistance.

3. Hysteresis effect: It was found that the measured voltage across the battery terminal is a function of previous current values that the battery experienced; this phenomenon is also known as the memory effect.

4. Relaxation effect: It was found that the terminal voltage of a battery gradually recovers to a higher value after discharging and to a lower value after charging; this phenomenon is known as the relaxation effect.

In the remainder of this section, we will discuss how electrical ECMs is used to represent the above four behaviors in real-world batteries.

3.3.2 Open-Circuit Voltage Model

The battery is a limited voltage source. The voltage across the terminals, denoted as EMF in Figure 3.6, drops as the battery is discharged and increases when the battery is charged. The state of charge of the battery is defined as

$$\text{SOC} = \frac{\text{remaining Coulombs (Ah)}}{\text{battery capacity (Ah)}} \tag{3.8}$$

where $\text{SOC} \in [0, 1]$ and the unit of the remaining Coulombs and the battery capacity are both in ampere hours (Ah). When the battery is charged, its SOC increases, and when it is discharged, its SOC decreases. The SOC of an empty battery is 0% and that of a fully charged battery is 100%. The SOC is sometimes displayed as a percentage in a typical electronic device, such as the one shown in Figure 3.7.

Figure 3.7 State of charge of a battery as displayed on a smartphone.

The EMF is referred to as the OCV of the battery. Theoretically, the OCV has a monotonous relationship with the SOC (i.e., as the SOC increases so does the OCV and vice versa). Figure 3.8 shows the OCV-SOC relationship of a typical Li-ion battery cell. When the battery is empty, the SOC is 0 or 0%, and when it is full, the SOC is 1% or 100%. The OCVs corresponding to an empty and full battery are denoted by OCV_{min} and OCV_{max}, respectively. The OCV-SOC relationship is useful to estimate the state of the charge of a battery. In simplified form, the SOC estimation approach works as follows: measure the OCV across the battery terminals, and then use the OCV-SOC curve in Figure 3.8 to look up the SOC of the battery. Given that the OCV-SOC curve is available as a parameterized function, the SOC estimation reduces to a root-finding problem.

One of the most important functions of a battery management system is SOC estimation. Hence, a BMS needs to have the parameters of the OCV-SOC curve. It is possible to derive the parameters of the OCV-SOC curve based on the exact chemical compositions of the battery cell. However, in practice, battery management systems use empirical approaches to obtain the OCV-SOC parameters. Chapter 4 provides details of various empirical approaches to estimate the parameters of an OCV-SOC curve.

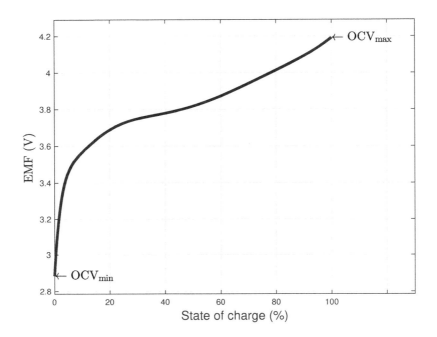

Figure 3.8 OCV-SOC model of a typical Li-ion battery.

Empirical OCV-SOC modeling approaches try to fit the measured OCV-SOC data using various functions. Chapter 4 provides an in-depth analysis of OCV-SOC modeling. Three sample OCV-SOC models are given below:

- Polynomial model

$$V_o(s) = p_0 + p_1 s + \cdots + p_n s^n \tag{3.9}$$

- Combined model

$$V_o(s) = \kappa_0 + \kappa_1 s^{-1} + \kappa_2 s + \kappa_3 \ln(s) + \kappa_4 \ln(1 - s) \tag{3.10}$$

- Combined+3 model

$$V_o(s) = k_0 + k_1 s^{-1} + k_2 s^{-2} + k_3 s^{-3} + k_4 s^{-4}$$
$$+ k_5 s + k_6 \ln(s) + k_7 \ln(1-s) \qquad (3.11)$$

where s denotes SOC and $V_o(s)$ denotes the corresponding OCV, that is,

$$\text{EMF} \equiv \text{OCV} \triangleq V_o(s) \qquad (3.12)$$

and the coefficients $\{p_0, p_1, \ldots, p_n\}$, $\{\kappa_0, \kappa_1, \ldots, \kappa_4\}$, and $\{k_0, k_1, \ldots, k_7\}$ are referred to as the OCV-SOC parameters of the polynomial, combined, and combined+3 models, respectively. In this book, the terms OCV-SOC parameters and OCV parameters are used interchangeably.

The selection of the OCV-SOC model, out of the three given in (3.9), (3.10), and (3.11) or from many other possible models for curve fitting, is left to the BMS designers. Most times, computational and hardware implementation constraints are prioritized over the accuracy of the model. Out of the three models presented, (3.9), (3.10), and (3.11), the combined+3 model is the most advanced in terms of accuracy. The combined+3 model is designed in a way to accommodate long linear portions and sudden declines usually found in OCV-SOC data. In this chapter, the combined+3 model is used in demonstrations.

Many existing OCV-SOC models, including (3.10) and (3.11), use terms such as $1/s, \log(s)$, and $\log(1-s)$, that are not defined at either $s = 0$ or at $s = 1$. In order to avoid numerical instability, a linear scaling approach can be employed. Figure 3.9 shows a linear scaling approach that can be used to avoid substituting $s = 0$ and $s = 1$ in (3.11). The scaling approach maps the SOC domain $s \in [0, 1]$ to $s' \in [0 + \epsilon, 1 - \epsilon]$ in a linear fashion as described in Figure 3.9. That way s' is prevented from reaching 0 or 1 and provides stability in computations [2]. Before computing the OCV, the SOC is scaled as

$$s' = (1 - 2\epsilon)s + \epsilon \qquad (3.13)$$

where the value of ϵ needs to be selected based on the model; it was reported in [2] that $\epsilon = 0.175$ gives optimal results in the combined model and its variants.

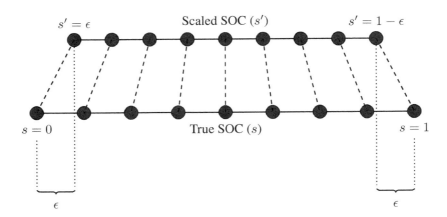

Figure 3.9 Linear scaling.

Example 3.2

The OCV parameters of a Samsung Galaxy S4 battery, according to the combined+3 model, are $k_0 = -9.082$, $k_1 = 103.087$, $k_2 = -18.185$, $k_3 = 2.062$, $k_4 = -0.102$, $k_5 = -76.604$, $k_6 = 141.199$, and $k_7 = -1.117$. These parameters were obtained after linearly scaling with $\epsilon = 0.175$. Use these parameters to create an OCV-SOC plot similar to the one shown in Figure 3.8.

An OCV-SOV plot corresponding to the given parameters can be produced through the following steps:

- Create SOC values spanning the entire region $[0, 1]$:

$$s = [0, 0.001, 0.002, \ldots, 0.999, 1] \tag{3.14}$$

- Scale SOC values in (3.14) using (3.13).
- For each of the scaled SOC, compute the corresponding OCV using (3.11) and the given parameters.
- Plot s (not s') against the computed OCV.

The resulting OCV-SOC plot will be very similar to the one shown in Figure 3.8. The following MATLAB code will produce an OCV-SOC plot using the parameters given in Example 3.2.

Listing 3.2: MATLAB Code for A Battery Simulator

```
clear all; clc; close all
% filename:SampleOCVSOCplot.m
% Creates OCV-SOC plot
% Uses Combined+3 model and scaling
k0 = -9.082; k1 = 103.087;
k2 = -18.185; k3 = 2.062;
k4 = -0.102; k5 = -76.604;
k6 = 141.199; k7 = -1.117;
SOC = 0:.0001:1;
epsilon = 0.175;
zs = SOC*(1-2*epsilon) + epsilon;
OCV = k0*ones(size(SOC)) ...
    + k1*(1./zs) + k2*(1./(zs.^2)) ...
    + k3*(1./(zs.^3)) + k4*(1./(zs.^4))...
    + k5*(zs) + k6*(log(zs))...
    + k7*(log(1-zs));
plot(SOC, OCV)
```

3.3.3 Relaxation Model

The relaxation effect observed in a battery is represented by the resistor and capacitor elements shown in Figure 3.3. The measured relaxation effect in a battery is shown in Figure 3.10 using real-world experimental data that was obtained as follows:

□

1. First 15 seconds: zero current.
   Because there is no current activity, the voltage is expected to remain constant during the first 15 seconds. Here, the battery was sufficiently rested in order to remove any prior relaxation effects. As expected, the measured voltage remained at 3.87V.

2. Next 30 seconds: constant discharge of $i_d = -1.171$A.
   When the current is applied, the voltage immediately dropped from 3.87V to 3.68V; this drop is due to the resistive elements in the ECM. Next the current remains constant and the OCV of the battery reduces as a result; this is observed through the continuous drop in the measured terminal voltage.

3. Final 135 seconds: zero current.
   When the discharge current is stopped, the voltage immediately returns to a higher value. After that, rather than remaining constant, the voltage keeps increasing. This observation demonstrates the presence of a relaxation effect within the battery.

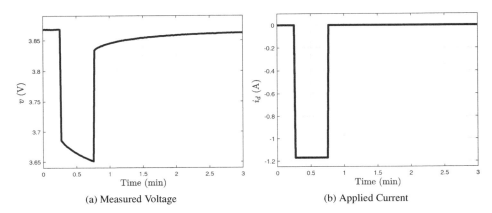

(a) Measured Voltage   (b) Applied Current

Figure 3.10 (a, b) Relaxation effect. The measured voltage across the battery in response to applied current over a 3-minute period.

Example 3.3

Consider the battery with the OCV parameters given in Example 3.2. The DC-ECM parameters of the battery are as follows: $R_0 = 0.2\Omega$, $R_1 = 0.1\Omega$, $C_1 = 2F$, $R_2 = 0.3\Omega$, $C_2 = 5F$. Consider the following 5-second load current profile that is applied to the battery: -40 mA from 0 to 1 second, -120 mA from 1 to 2 seconds, -120 mA from 2 to 3 seconds, -40 mA from 3 to 4 seconds, and -120 mA from 4 to 5 seconds. Assuming that the above current was sampled every $\Delta = 0.01$ second, compute the resulting voltage v across the battery terminals under the following model assumptions.

1. Ideal battery;

2. R-int model (for this case, assume R_1, R_2, C_1, C_2 are all zero);

3. RC model (for this case, assume R_2, C_2 are zero);

4. 2RC model.

Assume the same battery OCV parameters given in Example 3.2, that the battery is full at the start and that the battery capacity is $Q = 1.5$ Ah.

The answers to Example 3.3 are given below:

1. In an ideal battery, the terminal voltage is equal to $V_\circ(s(k))$ where $s(k)$ is the SOC at time k, which can be computed using the following Coulomb counting equation:

$$s(k+1) = s(k) + \frac{\Delta i(k)}{3600Q} \tag{3.15}$$

where $i(k)$ is the current through the battery in amperes. Once $s(k)$ is computed, the terminal voltage $v(k) = V_\circ(s(k))$ can be computed using (3.11).

2. Here, the terminal voltage is simply

$$v(k) = V_\circ(s(k)) + i(k)R_0 \tag{3.16}$$

3. In this case, the effect of the first RC element in Figure 3.3 will be included as follows:

$$v(k) = V_\circ(s(k)) + i(k)R_0 + i_1(k)R_1 \tag{3.17}$$

where $i_1(k)$ is the current through the resistor R_1 can be computed recursively using (3.2).

4. In this case, the effects of both of the first RC elements in Figure 3.3 will be included as follows:

$$v(k) = V_\circ(s(k)) + i(k)R_0 + i_1(k)R_1 + i_2(k)R_2 \tag{3.18}$$

where $i_2(k)$ is the current through the resistor R_2 can be computed recursively using (3.3).

The following MATLAB codes generate the answers to Example 3.3. It makes use of the battery simulator presented in Section 3.8.

Listing 3.3: MATLAB Code for Generating Voltage Across RC

```
clear all; clc; close all

k0 = -9.082; k1 = 103.087; k2 = -18.185;
k3 = 2.062; k4 = -0.102; k5 = -76.604;
k6 = 141.199;  k7 = -1.117;
Kbatt = [k0; k1; k2; k3; k4; k5; k6; k7];

Batt.Kbatt = Kbatt; Batt.Cbatt = 1.5;
Batt.R0 = .2; Batt.R1 = .1;
Batt.C1 = 2;  Batt.R2 = .3;
```

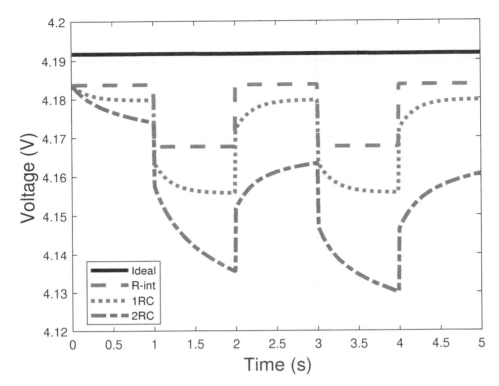

Figure 3.11 Reduced-order ECMs. Simulated terminal voltage for four different reduced-order ECMs in Example 3.3.

```
11 Batt.C2 = 5; Batt.ModelID = 4;
12 Batt.SOC = 1; Batt.SOCsf = 0.175;
13
14 % store current values in an array
15 delta = 0.01;
16 T = 0.01:delta:5;
17 I = -40*ones(1,length(T));
18 idx = find((T>1&T<2)|(T>3&T<4));
19 I(idx) = -120; I = I'/1000;
20
21 % Question 1 Plot
22 h1 = figure; box on; grid on; hold on
23 plot(T,1000*I,'LineWidth',2)
```

```
xlabel('Time (s)')
ylabel('Current (mA)')


% (1) Ideal battery model
Batt.ModelID = 2;
Batt.R0      = 0;
[V1] = battSIM(I, Batt, delta);
h=figure; box on; grid on; hold on
plot(T,V1,'LineWidth',3,'Color', [0 0 1])
% (2) R-int model
Batt.R0      = .2;
[V2] = battSIM(I, Batt, delta);
plot(T,V2,'LineWidth',3,'Color',[0 0.7 0], 'LineStyle','--')
% (3) 1RC model
Batt.ModelID = 3;
[V3] = battSIM(I, Batt, delta);
plot(T,V3,'LineWidth',3,'Color',[1 0.2 0.8], 'LineStyle',':')
% (4) 2RC model
Batt.ModelID = 4;
[V4] = battSIM(I, Batt, delta);
plot(T,V4,'LineWidth',3,'Color',[0 0.6 1], 'LineStyle','-.')

xlabel('Time (s)', 'fontsize', 15); ylabel('Voltage (V)', 'fontsize',
    15)
legend({'Ideal','R-int','1RC','2RC'},'location', 'best')
filename = '../../_figures/DCECMdemo';
print(h, '-depsc', filename)
```

☐

3.3.4 Hysteresis Model

The presence of hysteresis in a battery can be explained through the following observations:

- After applying a discharging current i_d, the SOC of the battery is computed to be s_1. Hence, the measured terminal voltage v is expected to return to $V_o(s_1)$ after the battery fully relaxes. However, in reality, the terminal voltage v returns to a value less than the expected $V_o(s_1)$.

- After applying a charging current i_c, the SOC of the battery is computed to be s_2. Hence, the measured terminal voltage v is expected to return to $V_o(s_2)$ after the battery fully relaxes. However, in reality, the terminal voltage v returns to a value higher than the expected $V_o(s_2)$.

  Experiments further concluded that:

- The magnitude of hysteresis voltage depends on the magnitude of current that it experienced over time (i.e., it is a function of both current and time).
- The magnitude of hysteresis voltage depended on the preceding SOC values of the batteries over time (i.e., it is a function of both SOC and time).

Based on experimental studies, the following model was presented in [3, 4] to explain the behavior of the hysteresis.

$$h(k+1) = \exp\left(-\left|\frac{\eta\gamma i_k \Delta_k}{Q}\right|\right)h(k) + \left(\exp\left(-\left|\frac{\eta\gamma i_k \Delta_k}{Q}\right|\right) - 1\right)\text{sgn}(i_k) \quad (3.19)$$

where k indicates time, η is the charging efficiency of the battery, Q is the battery capacity in Ah, γ, Δ_k is the sampling time, and i_k is the current (+ve for charging and -ve for discharging) through the battery.

Figure 3.12 shows a simulated hysteresis behavior in a battery based on the hysteresis model (3.19). In Figure 3.12(a), a battery was excited by two different current profiles, Profile A and Profile B, and the resulting hysteresis voltage is shown. Figure 3.12(b) shows the expected hysteresis when the profiles were switched in time. It shows the nonlinear relationship and the memory effect of hysteresis. In Figure 3.12(b), the same current profile, with a different time order, is used to simulate hysteresis. The resulting hysteresis voltage is found to be different from the one shown in Figure 3.12(a). This simulation explains the nonlinear and convoluted nature of the hysteresis effect in batteries.

To summarize, the more components the model has, the more accurate it becomes. However, increased model complexity implies increased difficulty, in terms of required data and computational complexity, in identifying the model parameters. Also, a certain type of data is needed to make some model parameters observable; this will be further illustrated later in Chapters 4, 5, and 6. In practical battery management systems, various ECM approximations are used. These approximations are generally referred to as reduced-order models. Some important reduced-order models are discussed in the next three sections.

3.3.5 Enhanced Self-Correcting Model

The enhanced self-correcting model was first introduced in [3] to represent the unknowns of a battery ECM in the form of a state-space model. The enhanced self-correcting model consists of the following process equation:

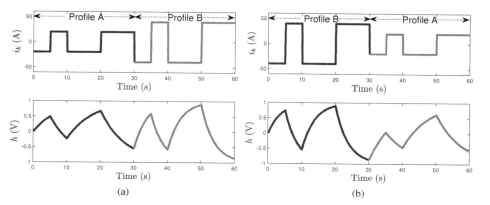

Figure 3.12 (a, b) Hysteresis voltage for two different current profiles. The magnitude of the hysteresis voltage is simulated and does not represent actual values in typical batteries.

$$
\begin{bmatrix} s(k+1) \\ i_1(k+1) \\ i_2(k+1) \\ h(k+1) \end{bmatrix} = \begin{bmatrix} 1 & 0 & 0 & 0 \\ 0 & \alpha_1 & 0 & 0 \\ 0 & 0 & \alpha_2 & 0 \\ 0 & 0 & 0 & A_h(k) \end{bmatrix} \begin{bmatrix} s(k) \\ i_1(k) \\ i_2(k) \\ h(k) \end{bmatrix}
$$
$$
+ \begin{bmatrix} A_c & 0 \\ 1-\alpha_1 & 0 \\ 1-\alpha_2 & 0 \\ 0 & A_h(k)-1 \end{bmatrix} \begin{bmatrix} i(k) \\ \mathrm{sgn}(i(k)) \end{bmatrix} \tag{3.20}
$$

and the following measurement equation

$$
v(k) = V_o(s(k)) + h(k) + i_1(k)R_1 + i(k)R_0 \tag{3.21}
$$

where $A_c = \frac{i_k \Delta_k}{Q}$ and $A_h(k) = \exp\left(-\left|\frac{\eta\gamma i_k \Delta_k}{Q}\right|\right)$.

Given voltage measurements $v(k)$, the enhance self-correcting model can be used to estimate the unknown quantities $s(k), i_1(k), i_2(k)$, and $h(k)$ assuming that the model parameters α_1, α_2, A_c, and $A_h(k)$ are known. When the model parameters are not known, advanced estimation techniques, such as the expectation-maximization algorithm, can be used to estimate them. More details about using the enhanced self-correcting model for battery state estimation are discussed in Chapter 8.

3.3.6   R-int Model

The R-int reduced-order battery ECM consists of the OCV-SOC model discussed in Section 3.3.2 and an internal resistance R_{int}.

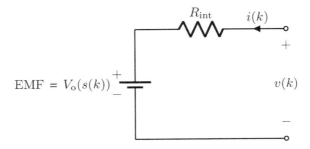

Figure 3.13 R-int model.

Using the R-int model, shown in Figure 3.13, the voltage across the battery terminals can now be written as

$$v(k) = V_o(s(k)) + i(k)R_{int} \tag{3.22}$$

where $i(k)$ is the current through the battery. It must be noted that the DC equivalent circuit model reduces to the R-int model when the current $i(k)$ remains constant for a long enough time while the hysteresis effect is ignored. Once the capacitors are saturated, one can write

$$R_{int} = R_0 + R_1 + R_2 \tag{3.23}$$

The R-int battery model is widely studied and adopted in BMS literature for its simplicity. In this book, various aspects of battery management will be discussed based on R-int with the understanding that the discussion can be generalized for ECMs that include both hysteresis and relaxation effects.

3.3.7 Other Reduced-Order Models

It is possible to derive many other reduced-order models from the DC-ECM shown in Figure 3.3. Indeed, the R-int model described in Section 3.3.6 is derived by ignoring the hysteresis and relaxation (RC) effects. An RC model is also widely used in the literature where one RC element instead of two is employed. In later chapters, several such reduced-order models will be used to explain various concepts. We will use these models with a slight adjustment of notation. For example, when an R-int model is used, the resistance will be denoted as R_0 instead of R_{int}. This convention is followed in this book to reduce the number of new notations for ECM parameters.

3.4 BATTERY POWER

The DC equivalent circuit model can be used to develop an expression of the available power in a battery. In this section, the R-int reduced-order model is considered for the derivations. Deriving an expression for the general DC-ECM is left as an exercise. The power of a battery at time k can be expressed using the following equation

$$P(k) = v(k)i(k) \tag{3.24}$$

where $P(k)$ represents the power at time k. When $i(k)$ is positive, the computed power denotes the input power to the battery, and when $i(k)$ is negative, the computed power denotes the output power.

Based on the R-int model, one can write

$$v(k) = V_o(s(k)) + i(k)R_0 \tag{3.25}$$

$$i(k) = \frac{v(k) - V_o(s(k))}{R_0} \tag{3.26}$$

Now the instantaneous power is written as

$$P(k) = v(k)\left(\frac{v(k) - V_o(s(k))}{R_0}\right) \tag{3.27}$$

Using the expression for instantaneous power in (3.27), the maximum allowable power of a battery during charging can be written as follows

$$P_{\text{In,max}}(k) = V_{\text{max}}\left(\frac{V_{\text{max}} - V_o(s(k))}{R_0}\right) \tag{3.28}$$

where $V_{max} = OCV_{max}$ is the maximum voltage allowed across the battery terminals. Similarly, the maximum available power during discharging can be written as

$$P_{Out,max}(k) = V_{min} \left(\frac{V_{min} - V_o(s(k))}{R_0} \right) \tag{3.29}$$

where $V_{min} = OCV_{min}$ is the minimum voltage allowed across the battery terminals. One can notice that $V_o(s(k)) > V_{min}$. Hence, maximum discharge power at a given time k can be written as $-P_{Out,max}(k)$.

Example 3.4

Consider the battery in Example 3.2.

1. What will be the open circuit voltage when the SOC of the battery is 60%?

2. Assuming the battery internal impedance of $R_0 = 0.2\Omega$, calculate the available peak power in the battery.

3. The battery was connected to a load that drew 300 mA constant-current for 1 hour. Assume that the battery capacity is 1.2 Ah. Compute the available peak power in the battery now.

4. Compute the open circuit voltage of the battery when it is full.

The answers to Example 3.4 are as follows:

1. To find the OCV at 60% SOC (or $s = 0.6$), we need to submit $s = 0.6$ in the combined+3 equation (3.11) for the given OCV parameters. Before that, scaling must be done as follows:

$$s' = (1 - 2\epsilon)s + \epsilon = (1 - 2\epsilon)0.6 + \epsilon = 0.565$$

Substituting $s' = 0.565$ in (3.11), we get

$$V_o(s) = OCV = 3.8723V$$

2. According to Figure 3.8, V_{min} occurs when $s = 0$. The scaled SOC is

$$s' = (1 - 2\epsilon)0 + \epsilon = 0.175$$

by substituting $s' = 0.175$ in (3.11), we get

$$V_{min} = V_o(s = 0) = 2.886$$

By substituting in (3.29),

$$P_{\text{Out,max}} = 2.886 \left(\frac{2.886 - 3.8723}{0.2} \right) \approx -14.23 \text{ W}$$

is the (maximum) available power
3. The new SOC of the battery can be computed as follows:
   - The initial SOC was $s = 0.6$.
   - The amount of Coulombs removed by the load is $0.3 \times 1 = 0.3$ Ah.
   - Removed Coulombs as a ratio of battery capacity is $0.3/1.2 = 0.25$.
   - New SOC is $0.6 - 0.25 = 0.35$.

The scaled version of the new SOC is

$$s' = (1 - 2\epsilon)s + \epsilon = (1 - 2\epsilon)0.35 + \epsilon = 0.4025 \tag{3.30}$$

Now, the OCV can be computed to be

$$V_o(s) = V_o(0.4025) = 4.1917 \text{ V} \tag{3.31}$$

By substituting in (3.29)

$$P_{\text{Out,max}} = 2.886 \left(\frac{3.78 - 2.886}{0.2} \right) \approx -12.9 \text{ W}$$

is the new available power
4. When the battery is full, the SOC becomes $s = 1$. The scaled version of the new SOC is

$$s' = (1 - 2\epsilon)1 + \epsilon = 0.825 \tag{3.32}$$

$$V_{\text{max}} = V_o(0.825) = 4.1917$$

□

3.5 BATTERY CAPACITY

Battery capacity is measured in ampere hours (Ah, i.e., a battery of x Ah in capacity can provide a load current of xA for one hour). This is only possible in theory. In practice, a battery of strictly x Ah in capacity cannot provide x A of load current for 1 hour due to the presence of internal resistance. In this section, we will discuss various ways to define battery capacity for practical applications.

Let us refer to the R-int equivalent circuit model shown in Figure 3.13 to understand more about the practical aspects of battery capacity. As indicated by the OCV-SOC curve in Figure 3.8, the EMF voltage of the battery reaches OCV_{max} when the battery is fully charged. As the charge is taken away from the battery, the EMF voltage drops, reaching OCV_{min} when the battery is fully empty. The change of EMF is not linear against the state of charge of the battery. It is also important to note that the operational region of the battery is between OCV_{min} and OCV_{max}, that is, the battery cannot be charged above OCV_{max} and neither can it be discharged below OCV_{min}. Charging a battery above OCV_{max} may cause thermal runaway, an irreversible process that triggers meltdown and fire. Discharging a battery below OCV_{min} may permanently damage the battery. These two constraints are the focus of several battery management functionalities discussed throughout this book. The cutoff voltages OCV_{max} and OCV_{min} are important factors to understand battery capacity.

The upper voltage threshold of the battery OCV_{max} serves as the charge cut-off voltage; that is, a battery charger monitors the voltage v at the battery terminal (see Figure 3.13) and terminates charging when $v \to OCV_{max}$. It must be noted that, due to the internal resistance R_0, the terminal voltage v will reach OCV_{max} before the EMF voltage reaches it. In other words, due to the internal resistance R_0, the battery charger must be switched off before the battery is completely full (which happens only when EMF reaches OCV_{max}). In order to fully charge a battery, typical Li-ion battery chargers switch to the constant-voltage mode in which the terminal voltage v is kept at a constant OCV_{max}. The charging current will gradually decrease to zero and the battery become fully charged. Figure 3.14 shows how the voltage, current, and SOC change during constant-current constant-voltage (CCCV) charging. Most practical Li-ion battery chargers employ CCCV topology.

The lower voltage threshold of the battery OCV_{min} serves as the discharge cutoff voltage (i.e., a BMS or protection circuit of a battery monitors the voltage v at the battery terminal (see Figure 3.13) and disconnects the load when $v \to OCV_{min}$). Similar to before, the internal resistance R_0 causes the terminal voltage v to reach OCV_{min} before the EMF voltage reaches it, (i.e., the battery needs to be shut down before the battery is completely empty). In other words, the available Coulombs of a battery depend on the voltage drop caused by the current and the internal resistance R_0. Due to this, different definitions of capacity are introduced next.

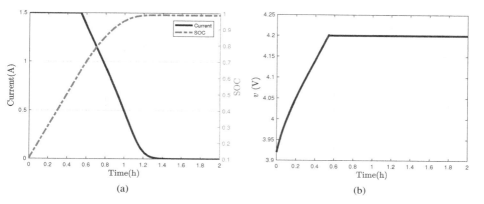

(a)   (b)

Figure 3.14 Constant-current constant-voltage (CC-CV) charging. (a) The current and SOC of the battery while charging. (b) The voltage of the battery while charging.

3.5.1 Total Capacity

Total battery capacity is the maximum amount of Coulombs that can be discharged from a battery, starting from a fully charged battery corresponding to its $OCV = OCV_{max}$ until the battery is fully discharged (i.e., until the battery terminal reaches $OCV = OCV_{min}$ using an infinitesimal load current). The total capacity specifies a theoretical value for the battery capacity. In order to achieve this, the charger switches to constant voltage mode and waits until (theoretically) the charging current becomes zero. In practice, battery capacity can be defined in many other ways to make useful interpretations. A few such definitions are demonstrated next.

3.5.2 Discharge Capacity

Discharge capacity indicates the amount of Coulombs that can be discharged from a battery at a certain discharge rate (load current). For example, consider a battery of 1.5 Ah in total capacity. Its discharge capacity could be 1.2 Ah at a 1A discharge rate or it could be 1 Ah at a 2A discharge rate. The drop in discharge capacity with higher load current is due to the fact that the voltage drop, which amounts to $i(k)R_0$ according to the R-int model shown in Figure 3.13, increases with the load current. As the voltage drop increases, the terminal voltage reaches the discharge cutoff before the battery is completely empty. Figure 3.15 describes this scenario where the discharge capacity is

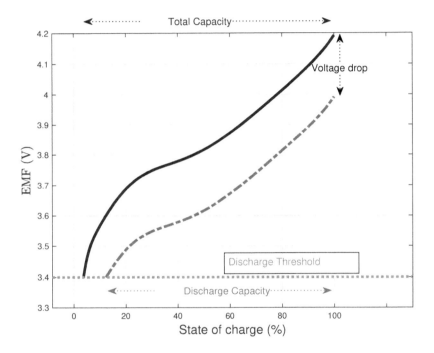

Figure 3.15 Total capacity versus discharge capacity in a battery.

less than the total capacity due to the voltage drop during discharging. As the voltage drop increases (i.e., as the load current increases), the discharge threshold OCV_{min} will be met earlier and the battery and the discharge capacity will be lower. The discharge capacity is also affected by temperature. In low temperatures, the internal resistance R_0 increases, resulting in higher voltage drop and lower discharge capacity.

3.5.3 Rated Capacity

Rated capacity is the manufacturer's specified discharge capacity of the battery. As the battery ages, the rated capacity becomes less accurate in representing the true discharge capacity of the battery. The rated capacity is accompanied by a discharge rate (e.g., 2 Ah at C-rate where C-rate is a popular way of specifying the load current in terms of battery capacity C).

Example 3.5

A battery of 10 Ah capacity is discharged at a $C/10$ rate. How long will it take to completely discharge the battery?

At $C/10$ rate, the discharge current is 1A; it will take 10 hours to completely discharge the battery. In general, when a battery is discharged at a C/N rate, it will last for N hours. □

3.5.4 Custom-Defined Capacity

It was described earlier that it takes infinite time to fully charge or discharge the battery. Coincidentally, it was discovered by researchers that completely charging the battery, or completely discharging it, has negative effects on battery health. Hence, it is beneficial to maintain the battery in such a way that its SOC is always maintained within a certain interval. For example, a battery whose SOC is always maintained between 30% and 70% will last longer and take less time to charge, compared to a battery whose SOC is maintained between 10% and 90%. Assume that the capacity of the above battery is 10 Ah; then the effective capacity of the battery is 4 Ah in the former case and 8 Ah in the latter case. That is, there is a trade-off involved in selecting an effective capacity; selecting a tighter SOC range results in a reduced rated capacity but a longer battery life and vice versa. It is also possible to select these SOC values to allow a certain load. This is elaborated using Example 3.6.

Example 3.6

A battery has 2 Ah in capacity and an internal resistance of $R_0 = 0.05\Omega$. It is required to charge this battery fast, using a constant current of 2A, and the highest load is also expected to be 2A.
(a) Design an effective SOC range that allows the above requirements.
(b) What is the effective capacity of the battery?
(c) How long will it take to fully charge the battery at the given rate?

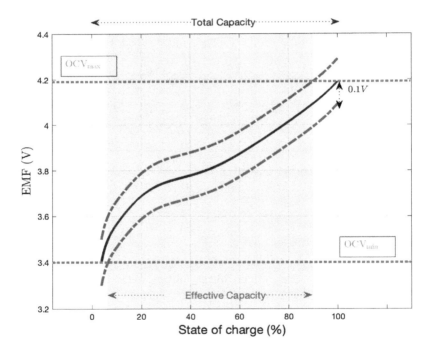

Figure 3.16 Custom battery design corresponding to Example 3.6.

Figure 3.16 shows the required voltage drop (of $0.05 * 2 = 0.1$V) during charging and discharging the battery. The battery needs to be shut off when the voltage drop hits the thresholds: OCV_{max} during charging and OCV_{min} during discharging and the corresponding SOC values form the required SOC range.

(a) According to a visual inspection in Figure 3.16, the SOC range is approximately $[5\%, 90\%]$.

(b) Hence, the effective capacity of the battery is 85% of the true capacity (i.e., 1.7 Ah).

(c) The battery, according to the above design, can be charged from empty to full at a constant rate of 2A. Hence, the charging time is

$$T_c = \frac{1.7\,\text{Ah}}{2\text{A}} = 0.85\,\text{hr} = 51\,\text{min}.$$

☐

In Example 3.6, the effective capacity is computed approximately by graphically drawing the voltage drop along with the OCV-SOC curve. Given the OCV-SOC parameters, the effective capacity can be precisely computed for given requirements.

Example 3.7

Consider the battery in Example 3.6. Assume that in a different application, the current requirement is increased to 4A. Recompute the effective SOC range, effective capacity and charging time.

Figure 3.17 shows the required voltage drop (of $0.05 * 4 = 0.2$V) during charging and discharging of the battery. The corresponding SOC values from the required SOC range are:

1. According to a visual inspection in Figure 3.17, the SOC range is approximately $[13\%, 78\%]$.
2. Hence, the effective capacity of the battery is 65% of the true capacity (i.e., 1.3 Ah).
3. The battery can be charged using a constant current of 2A. Hence, the charging time is

$$T_c = \frac{1.3\,\text{Ah}}{4\text{A}} = 0.325\,\text{hr} = 19.5\,\text{min}.$$

☐

3.6 STATE OF HEALTH

The SOH of the battery is defined in terms of power fade (PF) and capacity fade (CF). In this section, the formal definitions of PF and CF are defined first. Then approaches to defining SOH based on PF and CF are discussed.

Due to SEI growth and other internal chemical reactions the battery impedance increases over time. When the impedance of the battery increases, the output power

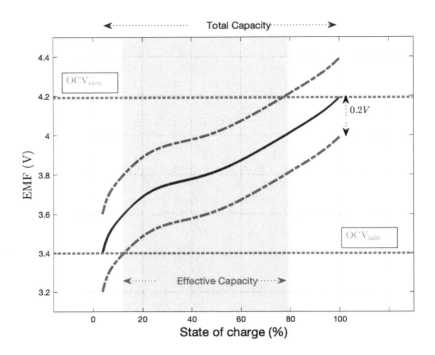

Figure 3.17 Custom battery design corresponding to Example 3.7.

decreases; this phenomenon is known as the power fade which is formally defined as

$$\mathrm{PF}(k) = \left(1 - \frac{P(k)}{P(0)}\right) 100\% \tag{3.33}$$

where the available power at time k is defined in (3.27). Here, the time index k indicates the elapse of life-cycle event, such as time and the charge-discharge cycle; it is assumed that $k = 0$ denotes the very initial cycle (e.g., brand-new battery). For accurate results, the power $P(k)$ must be computed at the same temperature and SOC.

As the battery ages, an increasing portion of its composition loses its ability to store energy resulting in reduced battery capacity over time. It must be noted that the battery CF starts from the time the battery is manufactured. Various factors such as deep discharge, full charge, and extreme temperature conditions aggravate the CF process.

The CF is formally defined as

$$\mathrm{CF}(k) = \left(1 - \frac{Q(k)}{Q(0)}\right) 100\ \%$$ (3.34)

where $Q(0)$ is the initial capacity of the battery and $Q(k)$ is the capacity at time k; similar to before, time k could be either a cycle number or a calendar time unit. Even though battery capacity does not fluctuate as widely as the ohmic resistance, small changes in capacity may occur against temperature. Hence, CF must be computed at the same temperature.

In practice, the exact values of the resistance $R_0(k)$ and capacity $Q(k)$ of the battery at time k are not known; they need to be estimated. Hence, the PF and CF equations need to be rewritten as

$$\mathrm{PF}(k) = \left(1 - \frac{\hat{P}(k)}{P(0)}\right) 100\ \%$$ (3.35)

$$\mathrm{CF}(k) = \left(1 - \frac{\hat{Q}(k)}{Q(0)}\right) 100\ \%$$ (3.36)

where $\hat{P}(k)$ and $\hat{Q}(k)$ are the estimated values of the power and capacity, respectively, at time k computed for a temperature.

The PF and CF measures can be unified into a single measure of SOH in various ways. Four such SOH definitions are:

1. $\mathrm{SOH} = 100 - \mathrm{PF}$
2. $\mathrm{SOH} = 100 - \mathrm{CF}$
3. $\mathrm{SOH} = 100 - \max\{\mathrm{PF}, \mathrm{CF}\}$
4. $\mathrm{SOH} = 100 - \min\{\mathrm{PF}, \mathrm{CF}\}$

The first definition above gives more priority to available power, whereas the second approach prioritizes battery capacity. The third and fourth definitions of the SOH are the most strictest and most relaxed, respectively.

All SOH measures introduced so far need to be computed based on the assumption that initial values (i.e., $P(0)$ and $Q(0)$) are known. The implication of this assumption is a significant limiting factor in present-day battery management systems. It remains challenging to evaluate the SOH of an arbitrary battery; such an ability will pave the way for safe, efficient, and reliable approaches to reuse batteries in various second-life applications.

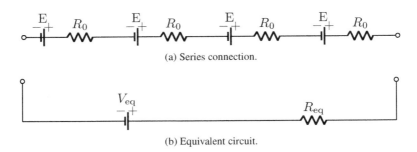

(a) Series connection.

(b) Equivalent circuit.

Figure 3.18 (a, b) Equivalent circuit model of series connection.

Remark 3.3 It is also important to be specific about how the capacity $Q(k)$ in (3.34) is computed. Many existing BMS productions (e.g., [5]) define SOH in terms of the rated capacity. Also, the user can set a particular discharge rate to define the SOH.

3.7 BATTERY PACKS

Battery packs are formed by connecting several battery cells in parallel and/or in series according to the power and capacity requirements. When identical battery cells are connected in series, the voltage across the terminal increases proportionally to the number of cells, whereas the battery capacity remains the same. Figure 3.18 shows the series connection of four identical battery cells. Here, the voltage, internal resistance, and capacity of the resulting pack are $V_{eq} = 4E$, $R_{0,eq} = 4R_0$, and $Q_{eq} = Q$ where Q is the capacity of the individual cell.

When identical battery cells are connected in parallel, the capacity of the resulting battery pack increases proportionally to the number of cells whereas the voltage remains the same. Figure 3.19 shows the parallel connection of four identical battery cells. Here, the voltage, internal resistance, and capacity of the resulting pack are $V_{eq} = E$, $R_{0,eq} = R_0/4$, and $Q_{eq} = 4Q$.

When battery cells are connected in parallel, the voltage across each cell in the pack is maintained at the same level by the natural flow of electrons. When the cells are connected in series, the voltage across each cell in the pack may not remain the same due to various factors. This phenomenon is known as cell imbalance, which may render the battery pack unusable over time. Chapter 9 motivates the need to balance a battery pack and discusses various ways of cell balancing.

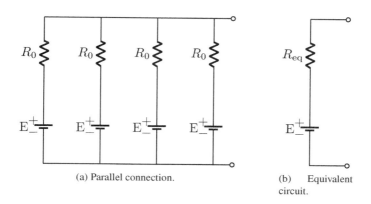

(a) Parallel connection.

(b) Equivalent circuit.

Figure 3.19 (a, b) Equivalent circuit model of parallel connection.

3.8 BATTERY SIMULATOR

Based on the DC-ECM discussed in this section, the following MATLAB codes can be used to simulate the voltage and current across a battery. Example 3.3 shows a use case of the following battery simulator.

Listing 3.4: **MATLAB Code for a Battery Simulator**

```
function [V, SOC, Vo] = battSIM(I, Batt, Delta)
    %% Reading battery ECM parameters
    Kbatt = Batt.Kbatt;
    Cbatt = Batt.Cbatt ;
    K0 = Kbatt(1); K1 = Kbatt(2); K2 = Kbatt(3);
    K3 = Kbatt(4); K4 = Kbatt(5); K5 = Kbatt(6);
    K6 = Kbatt(7); K7 = Kbatt(8);
    R0 = Batt.R0; R1 = Batt.R1; R2 = Batt.R2;
    C1 = Batt.C1; C2 = Batt.C2;
    ModelID = Batt.ModelID;
    E = Batt.SOCsf; % scaling factor
    alpha1=exp(-(Delta/(R1*C1)));
    alpha2=exp(-(Delta/(R2*C2)));
    %% Hysteresis model
    h = 0;
    %% Determine of SOC (Coulomb counting)
    SOC=zeros(length(I),1);
    SOC(1)=Batt.SOC; % initial SOC
    for k=2:length(I)
```

```
20          SOC(k)= SOC(k-1)+(1/(3600*Cbatt))...
21              *I(k)*Delta;
22          if SOC(k) < 0
23              error('Battery is Empty!!')
24          elseif SOC(k) > 1
25              error('Battery is Full!!')
26          end
27      end
28      %% Determination of OCV
29      zs = (1 - 2*E)*SOC + E;
30      Vo = K0+K1./zs + K2./(zs.^2) + ...
31          K3./(zs.^3)+K4./(zs.^4) + ...
32          K5*zs + K6*log(zs)+ K7*log(1-zs);
33      %% Determine current through R1 and R2
34      x1=zeros(length(I),1);
35      x2=zeros(length(I),1);
36      for k=1:length(I)
37      x1(k+1)=alpha1*x1(k)+(1-alpha1)*I(k);
38      x2(k+1)=alpha2*x2(k)+(1-alpha2)*I(k);
39      end
40      i1=zeros(length(I),1);
41      i2=zeros(length(I),1);
42      for k=1:length(I)
43          i1(k)=x1(k+1);  i2(k)=x2(k+1);
44      end
45      %% Determination of terminal voltage
46      V=zeros(length(I),1);
47      switch ModelID
48          case 1
49              V= I*R0;
50          case 2
51              V= I*R0+Vo+h;
52          case 3
53              V= I*R0+i1*R1+Vo+h;
54          case 4
55              V= I*R0+i1*R1+i2*R2+Vo+h;
56      end
57 end
```

3.9 SUMMARY

The goal of equivalent circuit model representation of a battery is to have the ability to represent important states of the battery, such as the state of charge, available power, and state of health, in terms of noninvasively and continuously measurable quantities. The electrical ECM consists of the following elements:

- Open circuit voltage model: The open circuit voltage of a battery has a monotonous and nonlinear relationship with the state of charge; as the state of charge increases, the OCV increases and vice versa. The OCV-SOC relationship is unique to each battery's chemistry, size, and physical shape. The nonlinear OCV-SOC relationship is required for battery management functionalities such as SOC estimation. For a given battery cell, the parameters of the OCV-SOC relationship are estimated in laboratories. This chapter has explained OCV parameter estimation.

- Series resistance: The series resistance of a battery changes with temperature and age; it also serves as a measure of battery age. Hence, accurate estimation of the series resistance is important for effective battery management.

- Relaxation effect: The relaxation effect is due to the response of a battery due to the capacitance along with the internal resistance.

- Hysteresis effect: The memory effect of the battery response is defined as hysteresis. Hysteresis modeling and its parameter identification remain a challenging research problem. Latter chapters will employ several techniques to account for the effect of hysteresis when estimating other battery parameters.

Each element of the ECM is defined by a set of parameters. These parameters can be estimated by carefully designed experiments. For example, the OCV parameters can be estimated by subjecting the battery to a carefully designed current profile. Chapter 4 presents approaches to estimate the OCV parameters of a battery. Chapter 5 details time-domain approaches to estimate the relaxation parameters of a battery, and Chapter 6 details frequency-domain approaches to estimate them. The parameters of the ECM may change with temperature, SOC, and age; a robust BMS needs to employ approaches to keep track of the ECM parameters at all times.

3.10 BIBLIOGRAPHICAL NOTES

The hysteresis model and the enhanced self-correcting model presented in this chapter are adopted from [3]. Some of the models presented in this chapter, such as the OCV and hysteresis models, were derived from empirical observations. The development of an OCV model directly based on the chemical composition of the cell is an active research area [6]. More details about the scaling approach presented in this chapter can be found in [2].

References

[1] C. Ho, I.D. Raistrick, and R.A. Huggins, "Application of AC techniques to the study of lithium diffusion in tungsten trioxide thin films," *Journal of the Electrochemical Society,* Vol. 127, No. 2, pp. 343, 1980.

[2] M.S. Ahmed, S.A. Raihan, and B. Balasingam, "A scaling approach for improved state of charge representation in rechargeable batteries," *Applied Energy*, Vol. 267, pp. 114880, 2020.

[3] G.L. Plett, *Battery Management Systems, Volume I: Battery Modeling*, Artech House, Norwood, MA, 2015.

[4] G.L. Plett, *Battery Management Systems, Volume II: Equivalent-Circuit Methods*, Artech House, Norwood, MA, 2015.

[5] Texas Instruments, "1-4 series Li-ion battery pack manager supporting Turbo Mode 2.0," serial number BQ40Z50-R2, https://www.ti.com/product/BQ40Z50-R2, (accessed Dec. 2021).

[6] C.R. Birkl, E. McTurk, M.R. Roberts, P.G. Bruce, and D.A. Howey, "A parametric open circuit voltage model for lithium ion batteries," *Journal of The Electrochemical Society* Vol. 162, No. 12, pp. A2271, 2015.

Chapter 4

Open-Circuit Voltage Characterization

4.1 INTRODUCTION

The open-circuit voltage (OCV) of a battery has a monotonically increasing relationship to its SOC. Using this relationship, the SOC can be computed by measuring the voltage across the battery terminals. This chapter explains how the OCV-SOC relationship is characterized and stored for realtime SOC estimation. This process is referred to as the OCV-SOC characterization or OCV characterization in this book.

The OCV characterization is done in laboratories using a scientific-grade battery cycler that is able to maintain precise voltage and current values across the battery terminals. Figure 4.1(a) shows a multichannel battery cell cycler made by Arbin Instruments [1]; this device allows the collection of battery characterization data simultaneously from many battery cells at the same time. The OCV-SOC characterization needs to be done at a fixed temperature. Figure 4.1(b) shows an environmental chamber made by Cincinnati Subzero for battery research; battery cells can be maintained at a fixed temperature in these chambers during the OCV test.

The data collection for the OCV characterization is designed in a way that the effects of the hysteresis and relaxation phenomenons of the battery can be nullified in the obtained OCV model. A fully charged battery is very slowly discharged (typically at a C/30 rate) using a constant current until it becomes empty. Then it is charged back to full charge using the same amount of constant current. This entire discharge-charge process takes 60 hours. Constant current assures that the capacitances of the equivalent circuit model remain saturated; a very low magnitude of current ensures that the hysteresis effect can be approximated as an equivalent resistance. By measuring the voltage and current values during this entire discharge/charge process, the OCV-SOC parameters

87

are obtained. It is preferred that this data is free of measurement noise and bias. High-precision battery cyclers can maintain constant currents with very little variation and can measure and store voltage and current with very little measurement noise.

(a) Battery cycler   (b) Environmental Chamber

Figure 4.1 (a, b) Scientific grade equipment used to collect OCV characterization data. Reproduced with permission from [2].

The OCV-SOC characterization is a curve fitting problem. The parameters of the OCV-SOC curve are obtained by fitting the collected data voltage-current data to a model; the current data is converted to SOC before the fitting. The parameters of this model are referred to as hereafter as the OCV parameters. BMS uses these parameters to look up SOC for a given OCV. The focus of this chapter is to introduce possible OCV-SOC models and approaches to fit the observed data.

Section 4.2 lists possible OCV models that have been used in the literature for OCV characterization. These models are classified under four categories: linear models, nonlinear models, hybrid models, and tabular models. Section 4.3 details the approach to estimating the parameters for the four types of models presented in Section 4.2. Given all the possibilities for OCV modeling, which model is suitable for a particular BMS design? Section 4.4 answers this question by introducing several model-selecting metrics. Finally, Section 4.5 provides an example of selecting a model under multiple constraints.

4.2 EMPIRICAL OCV-SOC MODELS

The open-circuit voltage of the battery has a monotonous relationship to the SOC. This relationship is the backbone of state-of-charge estimation algorithms. By measuring the voltage across the battery terminals, its SOC can be estimated as long as the OCV-SOC parameter is already available. In this section, we summarize some OCV-SOC functions reported in the literature.

4.2.1 Linear Regression Models

A linear OCV model can be written as

$$V_o(s) = \sum_{i=0}^{L-1} p_i(s)k_i = \mathbf{p}(s)^T \mathbf{k} \qquad (4.1)$$

where $V_o(s)$ denotes the OCV, $s \in [0, 1]$ denotes the SOC, and $\mathbf{p}(s)^T = [p_0(s), p_1(s), \ldots, p_{L-1}(s)]$ is a row vector of linear/nonlinear functions of s, and $\mathbf{k} = [k_0, k_1, \ldots, k_{L-1}]^T$ is the OCV parameter vector. The simplest form of the linear OCV model is the Unnewehr universal model, which is simply a straight line:

$$V_o(s) = \mathbf{p}(s)^T \mathbf{k} = k_0 + k_1 s \qquad (4.2)$$

which has $\mathbf{p}(s)^T = [1 \; s]$ and $\mathbf{k} = [k_0 \; k_1]^T$. Some linear models introduced in the literature are listed below:

- Shepherd model

$$V_o(s) = k_0 + \frac{k_1}{s} \qquad (4.3)$$

- Nernst model

$$V_\circ(s) = k_0 + k_1 \ln(s) + k_2 \ln(1 - s) \qquad (4.4)$$

- Combined model

$$V_\circ(s) = k_0 + \frac{k_1}{s} + k_2 s + k_3 \ln(s) k_4 \ln(1 - s) \qquad (4.5)$$

- Combined+3 model

$$V_\circ(s) = k_0 + \frac{k_1}{s} + \frac{k_2}{s^2} + \frac{k_3}{s^3} + \frac{k_4}{s^4} + k_6 s + k_6 \ln(s) + k_7 \ln(1 - s) \qquad (4.6)$$

- Polynomial model

$$V_\circ(s) = k_0 + k_1 s + \ldots + k_m s^m + k_{m+1} s^{-1} + \ldots + k_{m+n} s^{-n} \qquad (4.7)$$

- Exponential model

$$V_\circ(s) = k_0 + k_1 e^s + \ldots + k_m e^{s^m} + k_{m+1} e^{-s} + \ldots + k_{m+n} e^{-s^n} \qquad (4.8)$$

- Chebyshev model

$$V_\circ(s) = k_0 T_0(s) + k_1 T_1(s) + \ldots + k_{L-1} T_{L-1}(s) \qquad (4.9)$$

The parameters of all the linear models from (4.2) to (4.8) can be estimated through a linear least-square estimation approach. Section 4.3 demonstrates an example of linear least-squares model parameter estimation in detail. The parameters of the Chebyshev polynomial are computed as follows:

$$k_0 \quad = \quad \frac{1}{L} \sum_{k=1}^{L} V_\circ\left(\bar{s}_{ck}\right) \qquad (4.10)$$

$$k_i \quad = \quad \frac{2}{L} \sum_{k=1}^{L} V_\circ\left(\bar{s}_{ck}\right) T_i(\bar{s}_{ck}) \qquad (4.11)$$

where

$$\bar{s}_{ck} \quad = \quad \cos \frac{\pi\left(2k - 1\right)}{2L}, \quad k = 1, 2, ..., L \qquad (4.12)$$

$$T_i(\bar{s}_{ck}) \quad = \quad \cos \frac{i\left(2k - 1\right)\pi}{2L}, \quad i < L \qquad (4.13)$$

4.2.2   Nonlinear Regression Models

The nonlinear model is written in a general form as

$$V_\circ(s) = f(s, \mathbf{k}) \tag{4.14}$$

where $\mathbf{k}$ denotes a vector of model parameters. Similar to before, the number of parameters (i.e., the length of $\mathbf{k}$, depends on the model). Some possible nonlinear models from the literature are listed below:

- Double exponential model

$$V_\circ(s) = k_0 + k_1 s + k_2 \left(1 - e^{-k_3 s}\right) + k_4 \left(1 - e^{-\frac{k_5}{1-s}}\right) \tag{4.15}$$

- Nonlinear exponential model 1

$$V_\circ(s) = k_0 - \frac{k_1}{s} + k_2 e^{-k_3(1-s)} \tag{4.16}$$

- Nonlinear exponential model 2

$$V_\circ(s) = k_0 e^{-k_1 s} + k_2 + k_3 s + k_4 s^2 + k_5 s^3 \tag{4.17}$$

- Nonlinear exponential model 3

$$V_\circ(s) = a_1 e^{(b_1 s)} + a_2 e^{(b_2 s)} + c s^2 \tag{4.18}$$

- Rational approximant

$$V_\circ(s) = \frac{\sum_{i=0}^{m} k_i s^i}{1 + \sum_{j=1}^{n} k_{j+m} s^j}, m \geq 0, n > 0 \tag{4.19}$$

- Sum of sine model

$$V_\circ(s) = a_1 \sin(b_1 s + c_1) + a_2 \sin(b_2 s + c_2) + a_3 \sin(b_3 s + c_3) \tag{4.20}$$

4.2.3   Hybrid or Piecewise Linear Models

Hybrid modeling seeks to approximate the OCV-SOC curve as piecewise linear functions. One section of the OCV curve, where SOC $\in [0\ \zeta]$, is modeled using one of the linear functions presented in Section 4.2.1 and the other section of the OCV curve, where SOC $\in [\zeta\ 1]$, is modeled using another the linear function of Section 4.2.1. The advantage of hybrid modeling is that it offers better accuracy and computational efficiency in favor of more complex models. The formal representation of the two-piecewise linear OCV model is as follows

$$V_\circ(s) = \begin{cases} \mathbf{p}_i(s)^T \mathbf{k}_i & \text{if } s \geq \zeta \\ \mathbf{p}_j(s)^T \mathbf{k}_j & \text{if } s < \zeta \end{cases} \tag{4.21}$$

where each of $\mathbf{p}_i(s)$ and $\mathbf{p}_j(s)$ denote one of the several linear OCV models in Section 4.2.1. When $\mathbf{p}_i(s)$ and $\mathbf{p}_j(s)$ represent two straight lines, the minimum number of parameters will be five: two parameters each for the straight lines and ζ. These five parameters can be obtained using the hybrid optimization step briefed in Section 4.3.

4.2.4   Tabular Model

The OCV-SOC characterization approaches discussed so far required the estimation of a parameter vector $\mathbf{k}$. Depending on the model, the parameter $\mathbf{k}$ may require a high-precision floating point system for storage in a BMS. The tabular model stores the OCV and SOC pairs as a table. Table 4.1 shows an 11-point OCV-SOC table. The advantage of storing OCV-SOC characterization as a table is that it does not require a high-precision floating point system to store the values. The accuracy of the OCV-SOC table is not likely to be severely compromised by rounding these values (based on the available memory system). Some battery fuel gauging (BFG) algorithms require the derivative of the OCV function to recursively estimate SOC using filtering techniques. Hence, it is desired for an OCV-SOC table to store the derivatives as well.

Table 4.1 was formed by uniformly sampling SOC. The approximation error can be shown to be proportional to the curvature of the OCV-SOC curve. This implies a better strategy is needed to sample the OCV-SOC values for storage as a table. Section 4.3.4 details an improved approach to obtain samples for OCV-SOC tables.

4.3   OCV-SOC MODEL PARAMETER ESTIMATION

In this section, the detailed approach to estimating the OCV-SOC parameters, from data collection to parameter estimation, is presented. The data collection needs to be

Table 4.1

A Sample OCV-SOC Table

s	$V_o(s)$	$dV_o(s)/ds$
0.0	3.0519	44.4073
0.1	3.6594	0.6830
0.2	3.7167	0.8131
0.3	3.7611	0.4807
0.4	3.7915	0.4393
0.5	3.8275	0.6055
0.6	3.8772	0.8113
0.7	3.9401	0.9777
0.8	4.0128	1.0915
0.9	4.0923	1.1822
1.0	4.1797	1.3405

performed using professional, high-precision battery cyclers that have very low mea-surement noise. Figure 4.1(a) shows an Arbin battery cycler that can be programmed to execute the above data collection procedure. It is also important to keep the temperature fixed because the change in temperature translates to changes in internal resistance. A professional environmental chamber, similar to the one shown in Figure 4.1(b), needs to be used to keep the temperature fixed during the experiment.

The data for OCV characterization needs to be collected in a specific way such that the parameter estimation will not be affected by the elements of the equivalent circuit model in a battery. The following procedure needs to be followed for the data collection of OCV-SOC characterization:

1. Fully charge the battery at room temperature. In order to fully charge the battery, a constant-current (CC) constant-voltage (CV) approach can be used. The CV charging is terminated when the charging current i_c falls below $i_c < C/N$.

2. Bring the battery to a fixed temperature in which the OCV characterization is to be performed.

3. Slow-discharge the battery with a discharging current of $i_d = C/N$ rate until the terminal voltage reaches $v = \text{OCV}_{min}$. Let us denote the total discharge time as T_d.

4. Slow-charge the battery with a charging current of $i_c = C/N$ rate until the terminal voltage reaches $v = \text{OCV}_{max}$. Let us denote the total discharge time as T_c.

Here, the term C/N is used to indicate the magnitude of the current. For example, let us say that the manufacturer-rated capacity of the battery is $C = 1.5$ Ah. Then the current at $C/30$ rate is $i_c = i_d = 1.5/30 = 0.05$A.

The voltage and current data in the discharge and charge process are logged at a reasonable sampling rate. Considering that the discharge rate is very low, a sampling time of $\Delta = 60$ seconds is sufficient for OCV modeling. When N is set to $N = 30$ (i.e., when the discharging and charging currents are set to $i_c = i_d = C/30$A) the number of samples collected during discharging is $k_d = 30 \times 60 = 1800$. It must be noted that the actual number of k_d may vary depending on the available capacity of the battery, regardless of the labeled capacity.

Using the notations discussed so far, let us denote the voltage and current data collected during the discharging step (step 3) as $v(1)$, $v(2)$, ..., $v(k_d)$ and $i(1)$, $i(2)$, ..., $i(k_d)$, respectively. Similarly, let us denote the voltage and current data collected during the charging step (step 4) as $v(k_d + 1)$, $v(k_d + 2)$, ..., $v(k_c)$ and $i(k_d + 1)$, $i(k_d + 2)$, ..., $i(k_c)$, respectively. Using the notations introduced so far, the discharging and charging time can be written as

$$T_d = \Delta k_d, \quad T_c = \Delta(k_c - k_d) \tag{4.22}$$

The charge/discharge capacities of the battery are defined as

$$Q_c = i_c T_c, \quad Q_d = i_d T_d \tag{4.23}$$

where Q_c and Q_d denote the charge and discharge capacities, respectively.

Figure 4.2 shows the voltage $v(k)$ and current $i(k)$ measurements during the discharging and charging steps of the data collection from a sample battery. The data presented in this study is openly available in Mendeley Data at 10.17632/fywnpsjfpc.1, https://data.mendeley.com/datasets/fywnpsjfpc. The remainder of this section will detail how these data will be used to estimate the OCV parameters of a battery.

First, let us define the SOC at a given time as

$$s(k) \triangleq s \quad \text{at time } k \tag{4.24}$$

where the notation $\triangleq$, which reads "defined as," is used to assign a new variable name with a slightly different context; for example, the value s at time k is defined as $s(k)$ in (4.24).

The true SOC at time k can be recursively computed using the Coulomb counting equation:

$$s(k + 1) = s(k) + \frac{\Delta_k i(k)}{3,600Q} \tag{4.25}$$

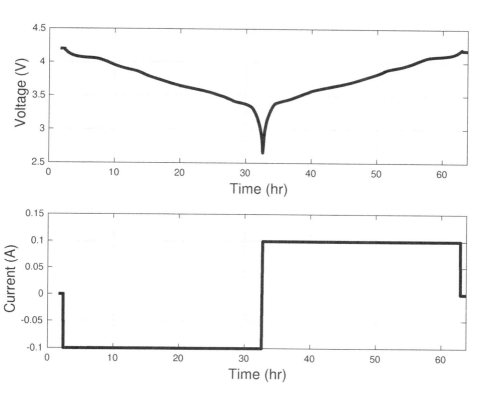

Figure 4.2 Measured voltage and current during charge and discharge at 25°C. Reproduced with permission from [2].

where $\Delta_k = \Delta$ is the time difference between two measurements in seconds, $i(k)$ is the current (in amperes) through the battery, and

$$Q = \begin{cases} Q_d & i(k) \leq 0 \\ Q_c & i(k) > 0 \end{cases} \tag{4.26}$$

is the battery capacity in Ah.

So far, $v(k)$ and $i(k)$ denoted the voltage across the battery terminals and current through the battery, respectively, during the experiment. Even in high-precision systems, the measured quantities will incur some measurement noise. Let us denote the measured voltage and current using the following

$$z_v(k) = v(k) + n_v(k) \tag{4.27}$$
$$z_i(k) = i(k) + n_i(k) \tag{4.28}$$

where $v(k), i(k)$ denote the true voltage across the battery terminals and current through the battery, $n_v(k)$ denotes the voltage measurement noise that is assumed to be zero-mean white with standard deviation (s.d.) σ_v, and $n_i(k)$ denotes the current measurement noise that is assumed to be zero-mean white with s.d. σ_i.

During the OCV experiment (i.e., when the battery is being slowly charged or discharged), the terminal voltage can be written as

$$z_v(k) = V_o(s(k)) + h(k) + i(k)R_{\text{int}} + n_v(k) \tag{4.29}$$

where $h(k)$ is the hysteresis voltage. By substituting $i(k) = z_i(k) - n_i(k)$ in (4.29), it can be rewritten in terms of the measured current as follows

$$z_v(k) = V_o(s(k)) + h(k) + z_i(k)R_{\text{int}} + n(k) \tag{4.30}$$

where the noise term $n(k)$ can be shown to be zero-mean with $\sigma_z = \sigma_v^2 + \sigma_i^2 R_{\text{int}}^2$ as the s.d.

It must be noted that the application of constant current throughout the entire experiment reduced the battery ECM to an R-int model described earlier. In terms of the regular time-domain ECM, the R-int can be written as

$$R_{\text{int}} = R_0 + R_1 + R_2 \tag{4.31}$$

Since the OCV test is performed at a very low current, it can be assumed that the hysteresis is proportional to the current only, that is,

$$h(k) \propto i(k) \tag{4.32}$$

Hence, (4.30) can be rewritten as

$$z_v(k) = V_o(s(k)) + z_i(k)R_{\text{eff}} + n(k) \tag{4.33}$$

where the effective resistance

$$R_{\text{eff}} = R_{\text{int}} + R_{\text{h}} \tag{4.34}$$

is the summation of the internal resistance R_{int} and the hysteresis equivalent resistance R_h.

The goal is to estimate the parameters that define the OCV $V_o(s(k))$ in (4.33). Depending on how the OCV is defined in Section 4.2, the parameter estimation approach needs to be different. For linear models summarized in Section 4.2.1, the linear least-squares method is explained in Section 4.3.1. For nonlinear models summarized in Section 4.2.2, the linear least-squares method is explained in Section 4.3.2. The parameter estimation of the hybrid linear models of Section 4.2.3 is summarized in Section 4.3.3, and Section 4.3.4 summarizes approaches to create OCV-SOC tables.

4.3.1 Linear Least-Squares

The linear OCV-SOC model parameter estimation approach is described in this section using one of the linear models, the combined+3 model (4.6), presented in Section 4.2.1. A similar approach can be followed to estimate all other linear models.

Using vector notations, the observation model in (4.33) can be rewritten as

$$z_v(k) = \underbrace{\left[\mathbf{p}_o(s(k))^T \quad i(k) \right]}_{\mathbf{p}(k)^T} \underbrace{\left[\begin{array}{c} \mathbf{k}_o \\ R_{\text{eff}} \end{array} \right]}_{\mathbf{k}} + n_v(k) \qquad (4.35)$$

where

$$\mathbf{k}_o = [k_0 \ k_1 \ k_2 \ k_3 \ k_4 \ k_5 \ k_6 \ k_7]^T \qquad (4.36)$$

and

$$\mathbf{p}_o(s(k))^T = [1 \ \frac{1}{s(k)} \ \frac{1}{s^2(k)} \ \frac{1}{s^3(k)} \ \frac{1}{s^4(k)} \ s(k) \ \ln(s(k)) \ \ln(1 - s(k))] \qquad (4.37)$$

By considering a batch of N voltage observations, (4.35) can be rewritten as

$$\mathbf{v} = \mathbf{P}\mathbf{k} + \mathbf{n} \qquad (4.38)$$

where

$$\mathbf{v} = [z_v[1], \ z_v[2], \ \ldots, \ z_v[k_d]]^T \qquad (4.39)$$

$$\mathbf{P} = [\mathbf{p}[1] \ \mathbf{p}[2] \ \ldots \ \mathbf{p}[k_d]]^T \qquad (4.40)$$

$$\mathbf{n} = [n[1] \ n[2] \ \ldots \ n[k_d]]^T \qquad (4.41)$$

$$\mathbf{k} = [k_0 \ k_1 \ k_2 \ k_3 \ k_4 \ k_5 \ k_6 \ k_7 \ R_{\text{eff}}]^T \qquad (4.42)$$

The least-squares estimate of the parameter vector is given by

$$\hat{\mathbf{k}} = \left(\mathbf{P}^T\mathbf{P}\right)^{-1}\mathbf{P}^T\mathbf{v} \qquad (4.43)$$

Now, for a given SOC, the corresponding OCV estimate $\hat{V}_\circ(s)$ is computed as

$$\hat{V}_\circ(s) = \mathbf{p}_\circ(s)^T\hat{\mathbf{k}}_\circ \qquad (4.44)$$

where $\hat{\mathbf{k}}_\circ$ is formed by the first 8 elements of $\hat{\mathbf{k}}$. Given the voltage and current data, $v(1), v(2), \ldots, v(k_c)$ and $i(1), i(2), \ldots, i(k_c)$, respectively, corresponding to the plot in Figure 4.2, the following MATLAB codes will generate the parameter vector $\mathbf{k}$ corresponding to the combined+3 model and generate Figure 4.3. This code can be copied from the Github site of the book
https://github.com/SingamLabs/Robust-Battery-Management-Systems.git under Chapter 4.

Listing 4.1: MATLAB Code for OCV Modeling

```
1 clear; clc; close all
2
3 %.mat file for battery data
4 data = importdata('C1204_OCV.mat');
5 id1  = [102:3894]; % to remove preceding charging
6 T    = data(id1,3)/3600;
7 I    = data(id1,7);
8 V    = data(id1,8);
9 idxc = find(I>=0);  % charge current
10 Ic = I(idxc);Tc = T(idxc);
11 idxd = find(I<0);  %  discharge current
12 Id = I(idxd); Td = T(idxd);
13
14 Qc = sum(diff(Tc).*Ic(2:end)); % charge capacity
15 Qd = -sum(diff(Td).*Id(2:end)); % discharge capacity
16
17 % compute SOC
18 SOC = zeros(length(I),1);
19 SOC(1) = 1; % Assumed that battery is full at start
20 for i = 2:length(I)
21     if I(i) >= 0
22         SOC(i) = SOC(i-1) + (T(i)-T(i-1))*(I(i))/(Qc);
23     else
24         SOC(i) = SOC(i-1) + (T(i)-T(i-1))*(I(i))/(Qd);
25     end
26 end
```

```
E   = .175;     % scaling factor
zs  = SOC*(1-2*E) + E; % scaled value of the SOC
N   = length(V);
P   = [ones(N,1) 1./zs 1./zs.^2 1./zs.^3 1./zs.^4 ...
        zs log(zs) log(1 -zs) I]; % combined model
kest = (P'*P)\(P'*V); % LS estimate
R0est = kest(end);
k   = kest(1:8);
OCV = k(1)*ones(length(zs),1) + k(2)*(1./zs) + k(3)*(1./(zs.^2)) + k(4)
     *(1./(zs.^3)) + ...
     + k(5)*(1./(zs.^4)) + k(6)*(zs) + k(7)*(log(zs)) + k(8)*(log(1-zs));

h=figure; hold on; grid on; box on; axis tight
plot(SOC,V, 'linewidth', 2, 'Color', [0 0 1])
plot(SOC,OCV, 'linewidth', 2, 'Color', [0 0.7 0],'LineStyle',':')
legend({'Measured Voltage', 'Modeled OCV'}, 'location', 'best')
xlabel('SOC', 'fontsize', 15);
ylabel('Voltage (V)', 'fontsize', 15);
filename = ['../_Ch04figures/4_3_ocvsocplot'];
print(h, '-depsc', filename)
```

The estimated OCV parameters are $k_0 = -6.6266$, $k_1 = 157.3029$, $k_2 = -26.8590$, $k_3 = 2.9721$, $k_4 = -0.1440$, $k_5 = -127.7601$, $k_6 = 224.5953$, $k_7 = -1.8463$, and $k_8 = 0.1984$; and the estimated effective resistance is $R_{\text{eff}} = 0.14809\Omega$.

4.3.2 Nonlinear Least-Squares

For nonlinear models, we rewrite (4.38) in the following form

$$\mathbf{v} = \mathbf{v}_o(\mathbf{k}_o) + \mathbf{i}R_{\text{eff}} + \mathbf{w} \tag{4.45}$$

where

$$\mathbf{k} = [\mathbf{k}_o \ R_{\text{eff}}]^T \tag{4.46}$$

$$\mathbf{v}_o(\mathbf{k}_o) = [V_o(s(1), \mathbf{k}_o) \ \dots \ V_o(s(k_c), \mathbf{k}_o)]^T \tag{4.47}$$

$$\mathbf{i} = [i(1) \ i(2) \ \dots \ i(k_c)] \tag{4.48}$$

and $\mathbf{w}$ is the noise vector.

The coefficients of the nonlinear regression-based models were computed using the MATLAB optimization toolbox function for nonlinear least-squares *lsqnonlin*. The nonlinear LS problem solves the following problem

$$\hat{\mathbf{k}} = \arg\min_{\mathbf{k}} \|\mathbf{v} - \hat{\mathbf{v}}\| \tag{4.49}$$

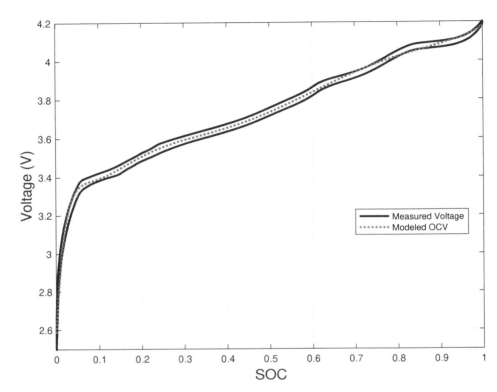

Figure 4.3 OCV model of a Li-ion battery. Reproduced with permission from [2].

4.3.3 Hybrid Estimation

Hybrid model parameters are estimated using constrained least-squares estimation techniques. The following constraints are used:

1. The derivative is always positive. This constraint ensures that the OCV is a monotonically increasing function in terms of SOC.

2. Both piecewise functions and their first and second derivatives are the same at the transition point ζ. This constraint ensures that the transition between one piecewise linear function to another is seamless and without any sudden changes in characteristics.

Let us rewrite the linear observation model (4.38) as

$$\begin{aligned}\mathbf{v}_1 &= \mathbf{P}^1(s)\mathbf{k}_1 + \mathbf{n}_1 \quad s \in [0, \zeta]\\ \mathbf{v}_2 &= \mathbf{P}^2(s)\mathbf{k}_2 + \mathbf{n}_2 \quad s \in [0, \zeta]\end{aligned} \tag{4.50}$$

The observation model (4.50) can be combined as follows

$$\begin{aligned}\tilde{\mathbf{v}} = \begin{bmatrix}\mathbf{v}_1\\ \mathbf{v}_2\end{bmatrix} &= \begin{bmatrix}\mathbf{P}^1(s) & 0\\ 0 & \mathbf{P}^2(s)\end{bmatrix}\begin{bmatrix}\mathbf{k}_1\\ \mathbf{k}_2\end{bmatrix} + \begin{bmatrix}\mathbf{n}_1\\ \mathbf{n}_2\end{bmatrix}\\ \tilde{\mathbf{v}} &= \tilde{\mathbf{P}}(s)\tilde{\mathbf{k}} + \tilde{\mathbf{n}}\end{aligned} \tag{4.51}$$

The model parameters of a hybrid, bilinear OCV-SOC function is obtained through the following optimization:

$$\{\hat{\mathbf{k}}_1, \hat{\mathbf{k}}_2\} = \arg \min_{\hat{\mathbf{k}}_1, \hat{\mathbf{k}}_2} \left\| \left(\tilde{\mathbf{P}}(s)\tilde{\mathbf{k}} - \tilde{\mathbf{v}} \right) \right\| \tag{4.52}$$

subject to

$$\frac{d\mathbf{P}^i\hat{\mathbf{k}}_1}{ds} > 0 \tag{4.53}$$

$$\frac{d\mathbf{P}^j\hat{\mathbf{k}}_2}{ds} > 0 \tag{4.54}$$

$$\mathbf{P}^i(s)\hat{\mathbf{k}}_1 \Big|_{s=\zeta} - \mathbf{P}^j(s)\hat{\mathbf{k}}_2 \Big|_{s=\zeta} = 0 \tag{4.55}$$

$$\frac{d\mathbf{P}^i(s)\hat{\mathbf{k}}_1}{ds}\Big|_{s=\zeta} - \frac{d\mathbf{P}^j(s)\hat{\mathbf{k}}_2}{ds}\Big|_{s=\zeta} = 0 \tag{4.56}$$

$$\frac{d^2\mathbf{P}^i(s)\hat{\mathbf{k}}_1}{ds^2}\Big|_{s=\zeta} - \frac{d^2\mathbf{P}^j(s)\hat{\mathbf{k}}_2}{ds^2}\Big|_{s=\zeta} = 0 \tag{4.57}$$

where $\|\cdot\|$ denotes the second norm.

The constrained least-squares solution *lsqlin* in the optimization toolbox of MAT-LAB can be used to solve the above optimization for a given value of ζ. The optimization can be repeated for a range of ζ values to find a better value for ζ that minimizes the cost function (4.52).

4.3.4 Tabular Model Estimation

In order to understand the need to have a better approach than uniform sampling, let us first define the approximation error. Consider a function $f(x)$ that is defined in $x \in [a, b]$. The goal is to represent this function at n discrete points, that is,

$$g(x) = \sum_{i=1}^{n} f(x)\delta(x - x_i) \quad i = 1, \ldots, n \qquad (4.58)$$

such that the sampling error is minimized. Let us define the sampling error as the following

$$e(x_i) = \frac{\Delta_i}{2}(f(x_i) + f(x_{i+1})) - \int_{x_i}^{x_{i+1}} f(x)dx \quad i = 2, \ldots, n \qquad (4.59)$$

where $\Delta_i = x_{i+1} - x_i$. The objective is to find a nonuniform sampling of the function such that the sum of the squared sampling errors in (4.59) is minimized. That is, for a given n,

$$\hat{\mathcal{X}} = \arg \min_{\mathcal{X}} \sum_{i=1}^{n} e(x_i)^2 \qquad (4.60)$$

where $\mathcal{X} = \{x_1, x_2, \ldots x_n\}$. Figure 4.4 shows an example of sampling error when $\Delta_i = x_{i+1} - x_i = \Delta$ (i.e., uniform sampling). It can be seen that the approximation error increases with the curvature (second derivative) of the function.

The curvature of the function $f(x)$ is formally defined as

$$C(x) = \frac{d^2 f(x)}{dx^2} \qquad (4.61)$$

Let us assume, without loss of generality, that the sign of curvature changes at $k (\geq 1)$ points and denote these k values as $x_{i_1}, x_{i_2}, \ldots, x_{i_k}$. That is, $x_i = \{x_{i_1}, x_{i_2}, \ldots, x_{i_k}\}$ will satisfy

$$C(x_{i_j}) = \frac{d^2 f(x)}{dx^2}\bigg|_{x=x_{i_j}} = 0, \quad j = 1, 2, \ldots, k \qquad (4.62)$$

These k values of x_i are denoted as the inflection points or critical points from now on. The nature of the curve significantly changes at critical points. When the function

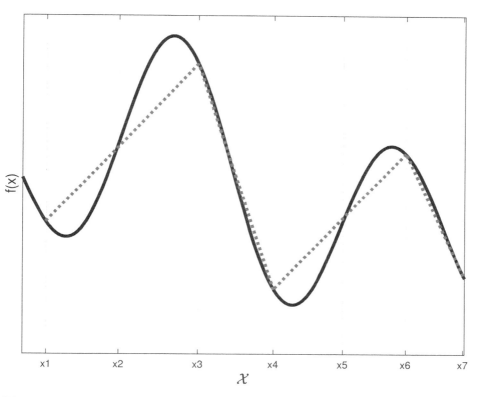

Figure 4.4 It can be seen that uniform sampling error increases with the magnitude of the curvature. Reproduced with permission from [2].

changes from convex to concave, the sign of the curvature changes from positive to negative, and vice versa; hence, one support point is assigned to each of the k critical points. Also, one support point is assigned, each at the start and end of the interval $[a, b]$, that is, out of the n available support points, $k + 2$ points, $x_1 = a$, $x_n = b$, and the k values of x_i are preassigned. These $k + 2$ points are denoted as the preassigned points from now on. This leaves us with $n - k - 2$ points to be assigned to $k + 1$ segments created by the k inflection points. The selection of these samples consists of the following two steps:

1. Find the number of support points to be allocated to each of the $k + 1$ sections created by the k inflection points.

2. Placement of support points in each section.

The following notations will be used to describe these approaches:

$$r = \left\lfloor \frac{n - k - 2}{k + 1} \right\rfloor \tag{4.63}$$

$$m = \mathrm{mod}\left(\frac{n - k - 2}{k + 1} \right) \tag{4.64}$$

where $\lfloor \cdot \rfloor$ denotes the floor operation and $\mathrm{mod}\,(\cdot)$ denotes the modulus operation. Here, one can see that

$$n - k - 2 = r(k + 1) + m \tag{4.65}$$

The absolute area of the curvature in each of the $k + 1$ sections is defined as

$$A_j = \int_{x_{i_j}}^{x_{i_{j+1}}} |C(x)|\, dx \quad j = 0, 1, \ldots, k \tag{4.66}$$

where $x_{i_0} = a$, $x_{i_{k+1}} = b$, and $x_{i_1}, x_{i_2}, \ldots, x_{i_k}$ are defined according to (4.62).

Remark 4.1 $x_{i_0} = a$ and $x_{i_{k+1}} = b$ are not the critical points.

Next, an approach is described to fulfill steps (1) and (2).

1. Number of support points:

- Each section gets r support points

- The remaining m support points are allocated as follows:

  If $m \leq 3$, the m support points are assigned to section j such that

$$A_j > A_i \quad \forall i = \{0, 1, \ldots, k\}, i \neq j \tag{4.67}$$

  ElseIf $m > 3$ and m is even, $m/2$ support points are assigned to each of section j_1 and section j_2 such that

$$A_{j_1}, A_{j_2} > A_i \quad \forall i = \{0, 1, \ldots, k\}, i \neq j_1, i \neq j_2 \tag{4.68}$$

  ElseIf $m > 3$ and m is odd, $\lceil m/2 \rceil$ support points are assigned to section j_1 such that

$$A_{j_1} > A_i \quad \forall i = \{0, 1, \ldots, k\}, i \neq j_1 \tag{4.69}$$

Else $\lfloor m/2 \rfloor$ support points are assigned to section j_2 such that

$$A_{j_2} > A_i \quad \forall i = \{0, 1, \ldots, k\}, i \neq j_1, i \neq j_2 \tag{4.70}$$

End

2. Placement of support points: Once the number of points in each section is allocated, the points in each section are then placed equally distant within that section. Let us assume that section j, which is bounded by x_{i_j} and $x_{i_{j+1}}$, was assigned L support points. The location of these L support points can be written as

$$x_l = x_{i_j} + d * l, \quad l = 1, 2, \ldots, L \tag{4.71}$$

where

$$d = \frac{x_{i_{j+1}} - x_{i_j}}{L + 1} \tag{4.72}$$

where the distance in each section is determined by the difference between the preassigned points of that section divided by the total number of points plus one of that section.

4.4 MODEL SELECTION METRICS

The OCV-SOC modeling approaches described in Section 4.2 are designed to minimize the mean square error. In this section, several other error metrics are introduced to assess the performance of an OCV-SOC model. A good model is expected to perform well across all error metrics introduced in this section. It is also important that an OCV model consists of as few parameters as possible. Several information theoretic metrics are introduced in this section to collectively evaluate models based on their error performance and the number of parameters that they require. In addition to this, there are other practical selection criteria for an OCV-SOC model: computational complexity, memory requirement, and numerical stability is important ones. This section provides brief discussions of these model selection criteria.

4.4.1 OCV Prediction Error

The following four error metrics can be used to evaluate OCV models.

1. Best-fit

$$BF(\%) = \left(1 - \frac{\|\hat{\mathbf{v}} - \mathbf{v}\|}{\|\mathbf{v} - \bar{\mathbf{v}}\|}\right) \times 100 \tag{4.73}$$

2. R^2 fit

$$R^2(\%) = \left(1 - \frac{\|\hat{\mathbf{v}} - \mathbf{v}\|^2}{\|\mathbf{v} - \bar{\mathbf{v}})\|^2}\right) \times 100 \tag{4.74}$$

3. Max-error

$$ME = \max_i \left\{|\mathbf{v}_i - \hat{\mathbf{v}}_i|\right\} \tag{4.75}$$

4. Root mean square error (RMSE)

$$RMS = \frac{\|\mathbf{v} - \hat{\mathbf{v}}\|}{\sqrt{N - M}} \text{ or } \sqrt{MSE} \tag{4.76}$$

where N is the number of data points, M is the number of parameters, $\hat{\mathbf{v}}$ is the predicted value of $\mathbf{v}$ using the estimated parameters, for example, for linear models

$$\hat{\mathbf{v}} = \mathbf{P}\hat{\mathbf{k}} \tag{4.77}$$

and

$$\bar{\mathbf{v}} = \frac{1}{N} \sum_{i=1}^{N} \hat{\mathbf{v}}(i) \tag{4.78}$$

The best-fit and R^2 fit metrics lie between 0 and 1; the higher the value, the better the model. For max-error and RMSE, the lower the value, the better the model.

4.4.2   Model Evaluation Metrics

Model evaluation metrics consider the trade-off between the number of model parameters and the number of data points. The following four metrics are important ones.

Table 4.2

OCV Prediction Errors

Model	BF	R^2	ME	RMSE
(4.2)	80.7150	96.2809	0.0490	0.0533
(4.3)	69.0215	90.4033	0.1806	0.0856
(4.4)	84.3122	97.5389	0.0785	0.0433
(4.5)	90.3010	99.0593	0.0839	0.0268
(4.6)	96.1031	99.8481	0.0544	0.0108
(4.7)	92.6999	99.4671	0.0562	0.0202
(4.8)	94.2150	99.6653	0.1055	0.0160
(4.15)	94.4633	99.6934	0.0445	0.0153
(4.16)	85.9646	98.0301	0.0904	0.0388
(4.17)	83.9417	97.4213	0.0827	0.0444
(4.18)	83.6839	97.3379	0.0806	0.0451
(4.19)	95.8975	99.8317	0.0455	0.0113
(4.20)	82.4227	96.9104	0.0622	0.0486
(4.71)	97.1593	99.9193	0.0640	0.0078

1. Akaike's Information Criterion-1: If the models are fitted using least-squares, then [3] suggests the following analog of AIC:

$$\text{AIC} = N \ln \left(\frac{S_2}{N} \right) + 2 \left(M + 1 \right) \tag{4.79}$$

where

$$S_2 = \sum_{i=1}^{N} e_i^2 \tag{4.80}$$

with

$$e = v - \hat{v} \tag{4.81}$$

In the above S_2 is the sum of the squares of errors (SSE), e_i is the ith element of the residual vector e and M is the number of parameters in the OCV model. The better the model, the lower the AIC.

2. Akaike's Information Criterion-2: A second version of AIC, given below, is useful when $N >> M$

$$\text{AIC2} = \ln\left[\mathcal{L}_f\left(1 + \frac{2p}{N}\right)\right] \tag{4.82}$$

where the loss function is defined as

$$\mathcal{L}_f = \frac{\mathbf{e}^T\mathbf{e}}{N} \tag{4.83}$$

3. Akaike's Final Prediction Error

$$\text{FPE} = \mathcal{L}_f\left[\frac{1 + \frac{M}{N}}{1 - \frac{M}{N}}\right] \tag{4.84}$$

4. Bayesian Information Criterion (BIC): The derivation of BIC assumes equal priors on each model and noninformative priors on the parameters, given each model. The goal of the BIC is to find the best (i.e., highest posterior probability) model for prediction.

$$\text{BIC} = 2\left(L_N\right) + (M + 1)\ln N \tag{4.85}$$

The negative log-likelihood, given the pdf of the residuals (assuming normal or Gaussian) conditioned on the parameters $\mathbf{k}$ and the s.d. of residuals σ is given by

$$L_N = -\ln\{L\left(\mathbf{k};\mathbf{e}\right)\} = \sum_{i=1}^{N}\left\{\left(\frac{\mathbf{e}_i^2}{2\sigma^2}\right) + 0.5\ln\left(2\pi\sigma^2\right)\right\} \tag{4.86}$$

where, L_N is the negative log-likelihood, L is the likelihood, $\mathbf{k}$ is the parameter vector which minimizes L_N, and σ is the s.d. of the residuals $\mathbf{e}$.

5. Minimum Description Length (MDL)

$$\text{MDL} = \mathcal{L}_f\left[1 + \frac{M\ln N}{N}\right] \tag{4.87}$$

Table 4.3

Model Evaluation Metrics

Model	AIC (e^3)	AIC2	FPE (e^{-3})	BIC(e^3)	MDL (e^{-3})
(4.2)	-2.1841	-5.8644	2.8403	-11.2449	2.8545
(4.3)	-1.8310	-4.9165	7.3290	-7.7138	7.3656
(4.4)	-2.3377	-6.2773	1.8805	-12.7746	1.8931
(4.5)	-2.6955	-7.2390	0.7196	-16.3406	0.7268
(4.6)	-3.3743	-9.0627	0.1163	-23.1092	0.1181
(4.7)	-2.9070	-7.8073	0.4079	-18.4492	0.4126
(4.8)	-3.0803	-8.2725	0.2561	-20.1822	0.2591
(4.15)	-3.1130	-8.3602	0.2346	-20.5090	0.2374
(4.16)	-2.4204	-6.4999	1.5060	-13.5956	1.5186
(4.17)	-2.3197	-6.2306	1.9736	-12.5761	1.9966
(4.18)	-2.3080	-6.1987	2.0364	-12.4657	2.0567
(4.19)	-3.3353	-8.9599	0.1292	-22.7015	0.1317
(4.20)	-2.2518	-6.0498	2.3684	-11.8781	2.4077
(4.71)	-3.6108	-9.7014	0.0617	-25.4266	0.0633

4.4.3 Computational Complexity

The OCV-SOC model, once created, is used by the BMS to estimate the SOC of the battery. That is, given the OCV, the BMS needs to use the OCV-SOC model parameters to compute the SOC. For most of the higher-order models, SOC estimation becomes a root-finding problem. For the Unnewehr linear model (4.2), SOC estimation becomes a closed-form equation, that is,

$$\hat{s} = \frac{z_v - k_0}{k_1} \qquad (4.88)$$

where z_v is a measure of the open-circuit voltage $(V_\circ(s))$ obtained by the BMS. Similarly, for tabular models, SOC estimation becomes a linear interpolation problem.

Table 4.4 compares all the models presented in this chapter in terms of their required computational complexity to find SOC for a given OCV. It is assumed that in all nonlinear root-finding cases, the bisection method is employed to find the SOC for a given OCV. It is also assumed that the bisection method uses 10 iterations in all cases. It is also assumed that all special functions are approximated for five terms using the Taylor series. The computational complexity shown in Table 4.4 refers to the number of additions (or subtractions) and multiplication (or divisions) needed to compute the SOC for a given OCV.

Table 4.4

Computational Complexity

Model	Complexity
(4.2)	1
(4.3)	1
(4.4)	110
(4.5)	130
(4.6)	160
(4.7)	60
(4.8)	260
(4.15)	120
(4.16)	70
(4.17)	90
(4.18)	110
(4.19)	1
(4.20)	150
(4.71)	16

4.4.4 Numerical Stability

Several OCV-SOC models employ functions such as e^s, $\ln(s)$, $\sin(s)$, and $\cos(s)$. Accurate implementation of these functions may require extra computing requirements that may not be affordable in some systems. Approximate implementations may result in errors. For example, the approximate implementation of $\ln(x)$ using Taylor series approximation can be written as

$$\ln(1-x) = -x - \frac{x^2}{2!} - \frac{x^3}{3!} - \frac{x^4}{4!} - \frac{x^5}{5!} - \cdots \tag{4.89}$$

An improved version of the above approximation of natural logarithm through Padé approximation:

$$\ln(1-x) \approx \frac{0.01812x^5 - 0.30555x^4 - 1.30555x^3 - 2x^2 + x}{0.00396x^5 - 0.11904x^4 + 0.83333x^3 - 2.22222x^2 + 2.5x - 1} \quad (4.90)$$

$$\approx \frac{((((137x - 2310)x + 9870)x - 15120)x + 7560)x}{((((30x - 900)x - 6300)x - 16800)x + 18900)x - 7560} \quad (4.91)$$

which can be implemented efficiently using Horner's method as follows:

$$\ln(1-x) \approx \frac{((((137x - 2310)x + 9870)x - 15120)x + 7560)x}{((((30x - 900)x - 6300)x - 16800)x + 18900)x - 7560} \quad (4.92)$$

Figure 4.5 compares three different implementations of $\ln(x)$: the first one is computed using high-precision computers (MATLAB), the second one is computed using (fifth order) Taylor series approximation (4.89), and the third one computed using the Padé approximation (4.92).

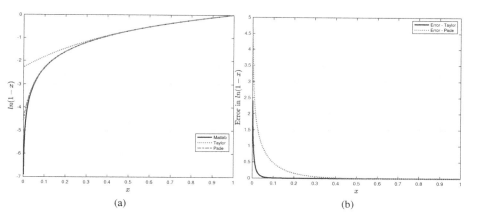

(a)   (b)

Figure 4.5 (a, b) Approximate implementation of the natural logarithm. Reproduced with permission from [2].

Figure 4.5 highlights the importance of numerical stability in the OCV-SOC models. The following two metrics are used to quantify the distortion of an OCV curve in relation to the original, high-precision form.

$$\text{KLD} = \sum_{i=0}^{n} \text{OCV}_0(s_i) \log \left(\frac{\text{OCV}_0(s_i)}{\text{OCV}_1(s_i)} \right) \qquad (4.93)$$

$$\text{CosD} = 1 - \left(\frac{\sum_{i=0}^{n} \text{OCV}_0(s_i)\text{OCV}_1(s_i)}{\sqrt{\sum_{i=0}^{n} \text{OCV}_0^2(s_i)} \sqrt{\sum_{j=0}^{n} \text{OCV}_1^2(s_i)}} \right) \qquad (4.94)$$

where KLD denotes the Kullback-Leiber divergence, CosD denotes the cosine distance, $\text{OCV}_0(s_i)$ denotes the OCV value computed by averaging the collected data for a given (discretized) SOC value s_i, and $\text{OCV}_1(s_i)$ denotes the OCV value computed by the model for the same SOC value s_i. The KLD and CosD metrics are computed for $n + 1$ different SOC values $s_0 = 0, s_1 = 1/n, s_2 = 2/n, \ldots, s_n = 1$ spanning the entire SOC range [0,1]. Table 4.5 summarizes the stability metrics computed for all the models presented in Section 4.2.

Table 4.5

Numerical Stability Metrics

Model	KLD (e^{-4})	CosD (e^{-4})
(4.2)	1.2328	1.0668
(4.3)	2.4222	2.3736
(4.4)	0.7993	0.6909
(4.5)	0.2592	0.2267
(4.6)	0.0283	0.0283
(4.7)	0.1367	0.1252
(4.8)	0.0705	0.0683
(4.15)	0.0691	0.0709
(4.16)	0.6226	0.5446
(4.17)	0.8277	0.7234
(4.18)	0.8577	0.7485
(4.19)	0.0253	0.0267
(4.20)	1.0064	0.8674
(4.71)	0.0639	0.0572

4.4.5 System Requirement

Once an OCV model is selected, its parameters need to be stored by the BMS for SOC estimation. In the case of combined model (4.5), the parameters $k_0, \ldots, k_4$ need to be stored. These parameters need to be selected in a way that the computational requirement is minimal. For example, consider the following values for the combined model parameters.

$$k_0 = -3.265420, k_1 = -1.090500, k_2 = 11.109784, k_3 = 2.972069, k_4 = -6.158655 \tag{4.95}$$

The parameters in (4.95) have six decimal points. The minimum system requirement to process SOC estimation using these parameters can be approximately stated as follows: 4 bits for the whole number, 20 bits for the fractional part, and 1 bit for sign, resulting in a total of 25 bits. In order to fit the above parameters to smaller systems, the parameters in (4.95) need to be rounded. Let us round the parameters to three decimal points:

$$k_0 = -3.265, k_1 = -1.090, k_2 = 11.110, k_3 = 2.972, k_4 = -6.159 \tag{4.96}$$

The system requirement to process these new sets of parameters in (4.96) is as follows: 4 bits for the whole number, 10 bits for the fractional part, and 1 bit for sign, resulting in a total of 15 bits.

Rounding the OCV parameters may result in SOC estimation errors. Figure 4.6 shows the effect of rounding in two different linear models. The rounding error is computed relative to the model that had parameters computed using MATLAB in a 64-bit system. Figure 4.7 shows the effect of rounding for a 16-point tabular approximation. It can be noticed from Figures 4.6 and 4.7 that the system requirement to achieve a certain level of maximum SOC error (e.g., 1% max. SOC error) varies from one model to another.

Table 4.6 summarizes the system requirement for each model presented in Section 4.2 in terms of the system required to maintain the maximum SOC error below 1%.

4.5 SELECTION OF THE OCV-SOC MODEL

The selection of the OCV-SOC model in practical situations is based on requirements that are specific to the application. For example, if high SOC estimation accuracy is required, then models with the lowest error metrics (Section 4.4.1) will be selected. This would imply that the computational and memory requirements are high. Most practical situations demand more than one constraint in model selection. The Borda count is an

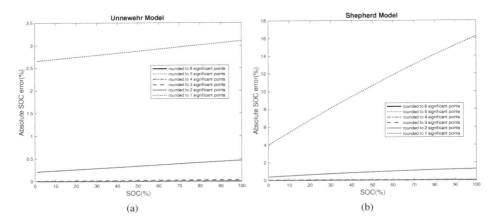

Figure 4.6 The effect of rounding in two different linear models: (a) Unnewehr and (b) Shepherd models. Reproduced with permission from [2].

intuitive method for combining different selection criteria for a compromised selection. The Borda count is originally a voting method in which each voter gives a complete ranking of all possible alternatives. Table 4.7 ranks the OCV models presented in this chapter based on all the selection criteria discussed in Section 4.4. In practical situations, one may choose to redo Table 4.7 based only on metrics that are important to the application.

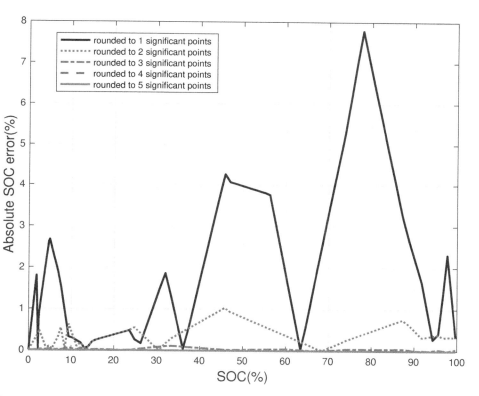

Figure 4.7 Effect of rounding. The effect of rounding in the tabular model is shown. Reproduced with permission from [2].

Table 4.6

System Requirements for OCV Parameters

Model	Parameters	Bits
(4.2)	$k_0 = 3.03, k_1 = 1.40$	12
(4.3)	$k_0 = 4.308, k_1 = -0.236$	15
(4.4)	$k_0 = 3.982, k_1 = 0.476, k_2 = -0.163$	15
(4.5)	$k_0 = -3.265, k_1 = --1.091, k_2 = 11.109,$ $k_4 = -6.158$	18
(4.6)	$k_0 = -6.62666, k_1 = 157.30292, k_2 = -26.85899,$ $k_3 = 2.97206, k_4 = -0.14404, k_5 = -127.76012,$ $k_6 = 224.59534, k_7 = -1.84633$	28
(4.7)	$k_0 = -9.8, k_1 = 27.4, k_2 = -24.9,$ $k_3 = 9, k_4 = -258.7$	15
(4.8)	$k_0 = 3.3522, k_1 = -0.8747, k_2 = 4.1821,$ $k_3 = -3.1107, k_4 = -0.0017, k_5 = -2.7797e - 15$	19
(4.15)	$k_0 = -228496.9, k_1 = 1.3,$ $k_2 = 492748.6, k_3 = 77.6,$ $k_4 = -264248.5, k_5 = 15$	22
(4.16)	$k_0 = 3.833, k_1 = 0.14,$ $k_2 = 0.866, k_3 = 2.822$	15
(4.17)	$k_0 = 1.024, k_1 = -0.355,$ $k_2 = 1.46, k_3 = 4.611,$ $k_4 = -6.898, k_5 = 4.016$	15
(4.18)	$a_1 = 1.308, a_2 = 1.307,$ $b_1 = 1.503, b_2 = 1.505, c = -7.140$	15
(4.19)	$k_0 = -47088.413, k_1 = 838478.954,$ $k_2 = 7077351.741, k_3 = 33622853.695,$ $k_4 = -77439664.427, k_5 = 68415176.728,$ $k_6 = 3058.908, k_7 = -379949.127,$ $k_8 = 4021098.125, k_9 = -12820987.612,$ $k_{10} = 13436120.396$	38
(4.20)	$a_1 = 3.3, a_2 = 3.3, a_3 = 1.76, b_1 = 1.14,$ $b_2 = 1.14, b_3 = 2.72, c_1 = 0.14, c_2 = 0.14,$ $c_3 = 2.11$	12
(4.71)	Rounded to 3 decimal points	8

Table 4.7
Model Selection Metrics Rankings

Model	BF	R^2	ME	RMSE	AIC	AIC2	FPE	BIC	MDL	KL	CosD	C	SR	Rank
(4.71)	1	1	7	1	1	1	1	1	1	3	3	2	1	1
(4.19)	3	3	2	3	3	3	3	3	3	1	1	1	8	2
(4.6)	2	2	4	2	2	2	2	2	2	2	2	10	7	3
(4.15)	4	4	1	4	4	4	4	4	4	4	5	7	6	4
(4.7)	6	6	5	6	6	6	6	6	6	6	6	3	3	5
(4.8)	5	5	13	5	5	5	5	5	5	5	4	11	5	6
(4.5)	7	7	11	7	7	7	7	7	7	7	7	8	4	7
(4.16)	8	8	12	8	8	8	8	8	8	8	8	4	3	8
(4.4)	9	9	8	9	9	9	9	9	9	9	9	6	3	9
(4.17)	10	10	10	10	10	10	10	10	10	10	10	5	3	10
(4.18)	11	11	9	11	11	11	11	11	11	11	11	6	3	11

4.6 SUMMARY

SOC estimation is one of the important tasks of a BMS. Coulomb counting is the simplest way to compute SOC. However, the Coulomb counting approach is dependent on two crucial pieces of information: battery capacity and initial SOC, both of which are significant uncertainties in a BMS. Voltage look-up-based SOC estimation does not need the knowledge of the battery capacity of the initial SOC. However, the voltage-based approach to SOC estimation needs to store the OCV-SOC curve in the form of its parameters. Obtaining the parameters of the OCV-SOC curve is known as OCV characterization. The OCV parameters are obtained from battery samples and stored in the battery management system. This chapter presented details about OCV-SOC characterization, from data collection to parameter estimation to storage. Various constraints that need to be considered before the selection of an OCV model are also presented in this chapter.

4.7 BIBLIOGRAPHICAL NOTES

Some of the discussions presented in this chapter are from [4]. The scaling approach is discussed and analyzed in detail in [5].

References

[1] Arbin website, https://arbin.com/.

[2] P. Pillai, S. Sundaresan, P. Kumar, K.R. Pattipati, and B. Balasingam, "Open-circuit voltage models for battery management systems: A review," *Energies,* Vol. 15, pp. 6803, 2022.

[3] K. P. Burnham, and D. R. Anderson, "Practical use of the information-theoretic approach," *Model Selection and Inference*, Springer, Berlin/Heidelberg, Germany, 1998.

[4] B. Pattipati, B. Balasingam, G.V. Avvari, K.R. Pattipati, and Y. Bar-Shalom, "Open-circuit voltage characterization of lithium-ion batteries," *Journal of Power Sources,* Vol. 269, pp. 317–333, 2014.

[5] M. Ahmed, S.A. Raihan, and B. Balasingam, "A scaling approach for improved state of charge representation in rechargeable batteries," *Applied Energy,* Vol. 267, No. 114880, 2020.

Chapter 5

Frequency-Domain Approaches to Battery ECM Identification

5.1 INTRODUCTION

Electrochemical impedance spectroscopy (EIS) [1, 2] is a widely used approach for battery analysis. In EIS, the frequency response of the battery, also referred to as the impedance spectrum, is studied for insights into battery health. Numerous works in the literature have employed impedance spectrums to understand and visualize battery aging. This chapter provides insights into the relationship between EIS and the ECM parameters of the battery; it is shown how to estimate the ECM parameters once the impedance spectrum is obtained. The estimated ECM parameters can then be related to SOH using the ideas provided in Chapter 3. This chapter also illustrates how parameter estimation becomes challenging with measurement noise in the sensors. Measurement noise is a significant problem in practical systems that tend to employ low-cost sensors; it is also possible that the measurement noise increases as the system ages. With measurement noise, the uncertainty about the (estimated) ECM parameters increases. It is important to have an understanding of the performance of ECM parameter estimation at various noise levels. This chapter introduces performance analysis of ECM parameter estimation, based on the Nyquist spectrum, at various signal-to-noise ratio (SNR) levels.

5.2 FREQUENCY RESPONSE OF A BATTERY

Randles circuit models are one of the widely used ECMs in frequency-domain analysis of batteries. Figure 5.1(a) shows a diagram of the adaptive Randles equivalent circuit model. This model consists of the following elements:

- Voltage source, E_{cell}

- Stray inductance, L

- Ohmic resistance, R_{Ω}

- Solid electrolyte interface (SEI) resistance, R_{SEI}

- SEI capacitance, C_{SEI}

- Charge transfer (CT) resistance, R_{CT}

- Double layer (DL) capacitance, C_{DL}

- Warburg impedance, Z_{w}

where the Warburg impedance is defined as

$$Z_{\text{w}}(j\omega) = (1 - j)\frac{\sigma}{\sqrt{\omega}} \tag{5.1}$$

and σ is referred to in this chapter as the Warburg coefficient. The Warburg impedance describes a phenomenon observed at very low frequencies. It can be noticed that as the frequency decreases, the Warburg impedance increases. When there is a significant frequency in the system, which is the case in practical applications, the Warburg impedance effectively becomes zero. However, at low frequencies, only the effects of resistors and the Warburg element remain.

In EIS, an AC perturbation signal (current or voltage) is applied to a battery and its response (voltage or current) is recorded; by jointly analyzing the applied signal and its response, the parameters of the battery ECM can be estimated. Figure 5.1(b) shows the typical response of a battery in the frequency domain; this impedance plot is generally known as the Nyquist plot, which shows negative imaginary part of the impedance $-\text{Im}(Z)$ against the real part of the impedance $\text{Re}(Z)$.

Using the AR-ECM shown in Figure 5.1, the AC impedance $Z(\omega)$ can be written as a function of angular frequency ω:

$$
\begin{aligned}
Z(\omega) &\triangleq Z(j\omega) \\
&= j\omega L + R_{\Omega} + \frac{1}{\frac{1}{R_{\text{SEI}}} + j\omega C_{\text{SEI}}} + \frac{1}{\frac{1}{R_{\text{CT}} + Z_{\text{w}}(j\omega)} + j\omega C_{\text{DL}}} \\
&= j\omega L + R_{\Omega} + \frac{R_{\text{SEI}}}{1 + j\omega R_{\text{SEI}} C_{\text{SEI}}} + \frac{R_{\text{CT}} + Z_{\text{w}}(j\omega)}{1 + j\omega \left(R_{\text{CT}} + Z_{\text{w}}(j\omega)\right) C_{\text{DL}}}
\end{aligned} \tag{5.2}
$$

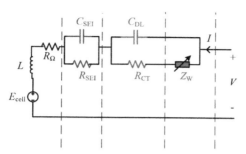

(a) Adaptive Randles equivalent circuit model (AR-ECM)

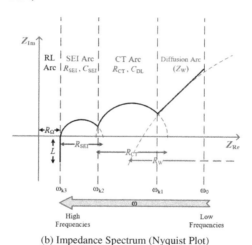

(b) Impedance Spectrum (Nyquist Plot)

Figure 5.1 (a) AR-ECM and (b) related impedance spectrum. Reproduced with permission from [3].

where the parameters are as indicated in Figure 5.1(a). Likewise, the qualitative impedance plot shown in Figure 5.1(b) has four branches associated with four specific electrochemical processes. In the first branch, denoted in this chapter as the RL Arc, the effect of the inductive behavior at high frequencies ($\omega > \omega_{k_3}$), as well as the ohmic resistance (R_Ω), can be seen. The second branch, ($\omega_{k_2} < \omega < \omega_{k_3}$), consists of a semicircle, denoted in this chapter as the SEI Arc, related to the solid electrolyte interface. The effect of the double-layer capacitance and charge transfer resistance at the electrodes represents the second semicircle (denoted as the CT Arc) in the third branch $(\omega_{k_1} < \omega < \omega_{k_2})$. Finally, the constant slope (denoted as the Diffusion Arc) in the impedance plot in the last branch represents the diffusion processes in the active material of the electrodes; it has a significant effect at very low frequencies $(\omega_0 < \omega < \omega_{k_1})$ only.

5.3 COMPUTING FREQUENCY RESPONSE USING DFT

The frequency response of the system, described in Section 5.2, can be computed using the discrete Fourier transform (DFT) technique. In order to compute the impedance spectrum, the voltage and current signals, measured in the time domain, are converted to the frequency domain through the DFT technique. Let us assume that $z_v(k)$ and $z_c(k)$ are the measured voltage and current, respectively, from the battery over a certain time window L, that is,

$$z_v(k) = v(k) + n_v(k), \quad k = 1, 2, \ldots, L$$
$$z_c(k) = i(k) + n_c(k), \quad k = 1, 2, \ldots, L$$
(5.3)

where k indicates time, $v(k)$ is the true voltage, and $i(k)$ is the true current. The voltage and current measurement noise $n_v(k)$ and $n_c(k)$ are assumed to be zero-mean i.i.d. with standard deviation σ_v and σ_c, respectively. The Fourier transforms of the voltage and current measurements in (5.3) are defined as

$$Z_c(\omega) = \text{FFT}(z_c(k)) = \sum_{k=1}^{L} z_c(k)e^{\frac{-i2\pi k\omega}{L}} = I(\omega) + N_c(\omega)$$
(5.4)

$$Z_v(\omega) = \text{FFT}(z_v(k)) = \sum_{k=1}^{L} z_v(k)e^{\frac{-i2\pi k\omega}{L}} = V(\omega) + N_v(\omega)$$
(5.5)

where $Z_v(\omega)$ and $Z_c(\omega)$ are the voltage and current measurements in the frequency domain. Here, $V(\omega)$ and $I(\omega)$ indicate the Fourier transforms of the noiseless voltage and current, respectively. The Fourier transforms of the voltage and current measurement noises are given by $N_v(\omega)$ and $N_c(\omega)$, respectively. The impedance at frequency ω is now written as

$$Z(\omega) = \frac{Z_v(\omega)}{Z_c(\omega)} = \frac{V(\omega) + N_v(\omega)}{I(\omega) + N_c(\omega)} = (V(\omega) + N_v(\omega))\left(\frac{1}{I(\omega) + N_c(\omega)}\right)$$
$$= (V(\omega) + N_v(\omega))\left(\frac{1}{I(\omega)} - \frac{N_c(\omega)}{I(\omega)^2}\right)$$
(5.6)

where the following Taylor series approximation was used

$$\left(\frac{1}{I(\omega) + N_c(\omega)}\right) \approx \frac{1}{I(\omega)} - \frac{N_c(\omega)}{I(\omega)^2} \tag{5.7}$$

Let us rewrite the impedance in (5.6) in the following format

$$Z(\omega) = \frac{V(\omega)}{I(\omega)} + N_z(\omega) \tag{5.8}$$

where it can be shown that the noise $N_z(\omega)$ is zero-mean. Let us denote the real and imaginary parts of the frequency response at ω_k as

$$z_r(k) = z_r(\omega_k) = \text{Re}(Z(\omega_k))$$
$$z_i(k) = z_i(\omega_k) = \text{Im}(Z(\omega_k)) \tag{5.9}$$

We will make use of this notation to describe the ECM parameter estimation approaches in the subsequent sections.

Remark 5.1 It should be noted here that the Fourier analysis assumes no transient effect for the excitation signal. The data collection must be done in a way that the transient effects will be minimal. If the transient effect is suspected, then Laplace analysis needs to be used instead of Fourier.

5.4 ECM PARAMETER ESTIMATION PROBLEM

The ECM parameter estimation in the frequency domain can be formally stated as follows: given the frequency response of the system $Z(\omega)$ at the frequencies $\omega_1, \omega_2, \ldots, \omega_L$, estimate the ECM parameters.

$$\hat{\Theta} = \arg\min_{\Theta} \left\| Z(\omega_i) - \hat{Z}(\omega_i) \right\|, \quad i = 1, \ldots, L \tag{5.10}$$

where $\|\cdot\|$ denotes two-norm,

$$\Theta = \{R_\Omega, L, R_{\text{SEC}}, C_{\text{SEC}}, R_{\text{CT}}, C_{\text{DL}}, Z_W\} \tag{5.11}$$

An approximate solution to the above optimization problem can be obtained using a nonlinear least squares approach.

5.5   APPROXIMATE ESTIMATION OF ECM PARAMETERS

The geometric features of the impedance spectrum can be exploited to compute the parameters of the Randles ECM. In this section, an approximate method is explained.

Consider the impedance in (5.2) at very high frequencies $\omega > \omega_3$. The capacitive reactance approaches zero at very high frequencies and only the inductive and ohmic resistances remain dominant. Hence, one can write

$$Z(\omega) \approx j\omega L + R_\Omega \quad \omega > \omega_{k_3} \tag{5.12}$$

From this, the estimates of L and R_Ω can be obtained as

$$\hat{R}_\Omega = \min\left(z_r(k)\right) \quad k > k_3 \tag{5.13}$$

$$\hat{L} = \frac{\left|\min\left(-z_i(k)\right)\right|}{\omega} \quad k > k_3 \tag{5.14}$$

At very low frequencies, the Warburg impedance (5.1) becomes dominant. The resistive part of the Warburg impedance is written as

$$R_\mathrm{w} = \frac{\sigma}{\sqrt{\omega}} \tag{5.15}$$

Consequently, the real part of $Z(\omega)$, when $\omega < \omega_{k_1}$, can be written as

$$z_r(k) \approx R_\Omega + R_\mathrm{SEI} + R_\mathrm{CT} + R_\mathrm{w} \quad k < k_1 \tag{5.16}$$

From (5.16), σ can be calculated as the slope of $z_r(k)$ versus $\frac{1}{\sqrt{\omega_k}}$ such that $k < k_1$. To compute σ, two low frequencies are chosen and the corresponding resistance value is taken from the impedance plot. Let us select these two frequencies as follows

$$\omega_a = \omega_0 \tag{5.17}$$

$$\omega_b = \omega \quad \text{s.t. } \omega_0 < \omega < \omega_{k_1} \tag{5.18}$$

Then the Warburg coefficient can be written as follows:

$$\hat{\sigma} = \frac{(\sqrt{\omega_a \omega_b})(z_r(a) - z_r(b))}{\sqrt{\omega_a} - \sqrt{\omega_b}} \tag{5.19}$$

Now let us consider the two arcs (SEI Arc and CT Arc) in the Nyquist plot (or impedance spectrum) to determine the value of R_SEI, C_SEI, R_CT, and C_DL. First, consider the CT Arc, which occurs in lower frequencies (i.e., $\omega_{k_1} < \omega < \omega_{k_2}$). The (Faradaic) impedance due to R_{CT} and C_{DL}, in this region is

$$Z_{\mathrm{F}}(\omega) = \cfrac{1}{\frac{1}{R_{\mathrm{CT}}+Z_{\mathrm{w}}(j\omega)} + j\omega C_{\mathrm{DL}}}$$

$$= \cfrac{\frac{1}{R_{\mathrm{CT}}+Z_{\mathrm{w}}(j\omega)} - j\omega C_{\mathrm{DL}}}{\left(\frac{1}{R_{\mathrm{CT}}+Z_{\mathrm{w}}(j\omega)}\right)^2 + \omega^2 C^2_{\mathrm{DL}}} \qquad \omega_{k_1} < \omega < \omega_{k_2} \qquad (5.20)$$

By ignoring the effect of Warburg impedance, the impedance corresponding to the CT Arc can be written as

$$Z_{\mathrm{CT}}(\omega) \approx \cfrac{\frac{1}{R_{\mathrm{CT}}} - j\omega C_{\mathrm{DL}}}{\left(\frac{1}{R_{\mathrm{CT}}}\right)^2 + \omega^2 C^2_{\mathrm{DL}}} \qquad \omega_{k_1} < \omega < \omega_{k_2} \qquad (5.21)$$

At the peak of CT Arc, one can observe (see Figure 5.1(b))

$$\frac{1}{\omega C_{\mathrm{DL}}} = R_{\mathrm{CT}} \quad \text{at } \omega = \omega_{\mathrm{CT,peak}} \qquad (5.22)$$

$$|\mathrm{Im}(Z_{\mathrm{CT}}(\omega))| = \frac{1}{2}R_{\mathrm{CT}} \quad \text{at } \omega = \omega_{\mathrm{CT,peak}} \qquad (5.23)$$

and the following two estimates can be obtained:

$$\hat{R}_{\mathrm{CT}} = 2|\mathrm{Im}(Z_{\mathrm{CT}}(\omega))| \quad \text{at } \omega = 2\pi f_{\mathrm{CT,peak}} \qquad (5.24)$$

$$\hat{C}_{\mathrm{DL}} = \frac{1}{\omega \hat{R}_{\mathrm{CT}}} \quad \text{at } \omega = 2\pi f_{\mathrm{CT,peak}} \qquad (5.25)$$

where $f_{\mathrm{CT,peak}}$ denotes the frequency corresponding to the peak of the CT Arc.

Let us now consider the SEI Arc in the range of $\omega_{k_2} < \omega < \omega_{k_3}$. In this region, the impedance is given by

$$Z_{\mathrm{SEI}}(\omega) = \cfrac{1}{\frac{1}{R_{\mathrm{SEI}}} + j\omega C_{\mathrm{SEI}}} = \cfrac{\frac{1}{R_{\mathrm{SEI}}} - j\omega C_{\mathrm{SEI}}}{\frac{1}{R^2_{\mathrm{SEI}}} + \omega^2 C^2_{\mathrm{SEI}}} \qquad \omega_{k_2} < \omega < \omega_{k_3} \qquad (5.26)$$

At the peak of SEI Arc, we have (see Figure 5.1(b))

$$\frac{1}{\omega C_{\mathrm{SEI}}} = R_{\mathrm{SEI}} \quad \text{at } \omega = \omega_{\mathrm{SEI,peak}} \qquad (5.27)$$

Based on the above observation, we have

$$|\text{Im}(Z_{\text{SEI}}(\omega))| = \frac{1}{2}R_{\text{SEI}} \quad \text{at } \omega = \omega_{\text{SEI,peak}} \tag{5.28}$$

and the following two estimates can be obtained:

$$\hat{R}_{\text{SEI}} = 2|\text{Im}(Z_{\text{SEI}}(\omega))| \quad \text{at } \omega = 2\pi f_{\text{SEI,peak}} \tag{5.29}$$

$$\hat{C}_{\text{SEI}} = \frac{1}{\omega \hat{R}_{\text{SEI}}} \quad \text{at } \omega = 2\pi f_{\text{SEI,peak}} \tag{5.30}$$

where $f_{\text{SEI,peak}}$ denotes the frequency corresponding to the peak of the SEI Arc.

Despite its simplicity and low processing time, the algorithm reviewed in this section has some drawbacks due to the effect of approximation and the effect of measurement noise. These effects are described next.

5.6   CAUSES OF PARAMETER ESTIMATION ERROR

This section explains some of the causes of errors in the approximate ECM parameter estimation approach summarized in Section 5.4.

5.6.1   Effect of Approximation

For the approximation parameter estimation presented in this section, the peak values of SEI Arc and CT Arc are found and the value of RC elements are calculated using (5.23) to (5.30). However, it should be noted that, for these calculations, the effect of the estimated Warburg impedance on CT Arc is neglected. Figure 5.2 illustrates how this approximation will affect the parameter estimation. For the case shown in Figure 5.2, the frequency at which the CT Arc reaches its maximum value is changed by more than 50% (from 0.59 to 0.91); that will affect the values of R_{CT} and C_{DL} according to (5.24).

5.6.2   Effect of Measurement Noise

The measurement noise also plays a major role in the accuracy of estimated ECM parameters. In order to analyze the performance, let us first define the SNR as

$$\text{SNR} = 10 \log \left(\frac{P_{\text{Signal}}}{P_{\text{Noise}}} \right) \tag{5.31}$$

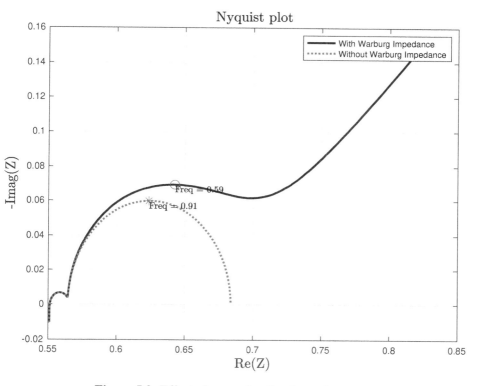

Figure 5.2 Effect of approximation in performance.

where the unit of SNR is decibels (dB). Figure 5.3 shows the impedance spectrum at four different SNR values (0, 5, 15, 30 dB). The algorithm summarized in Section 5.5 is applied to estimate the AR-ECM parameters in each case. The estimated parameters are then used to generate the impedance spectrum. Ideally, both plots should coincide. It can be noticed that, with increasing noise, the discrepancies become prominent.

5.7   IMPROVED APPROACH FOR PARAMETER ESTIMATION

Let us denote the frequencies at different branches of the impedance spectrum as follows:

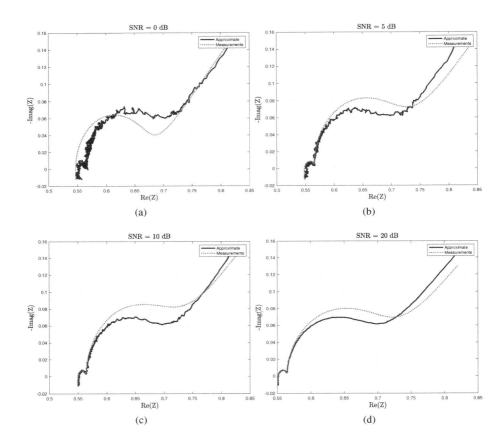

Figure 5.3 (a–d) Effect of measurement noise on performance.

- Warburg: $\omega_0, \omega_1, \ldots, \omega_{k_1}$

- CT: $\omega_{k_1+1}, \omega_{k_1+2}, \ldots, \omega_{k_2}$

- SEI: $\omega_{k_2+1}, \omega_{k_2+2}, \ldots, \omega_{k_3}$

- RL: $\omega_{k_3+1}, \omega_{k_3+2}, \ldots, \omega_{k_4}$

In total, it is assumed that there are k_4 frequency pairs at which the impedance measurements are to be computed. Let us denote the real and imaginary parts of the impedance measurements at each of the above frequencies as follows:

• Warburg:

$$[z_r(0), z_i(0)], [z_r(1), z_i(1)], \ldots, [z_r(k_1), z_i(k_1)] \tag{5.32}$$

• CT:

$$[z_r(k_1+1), z_i(k_1+1)], [z_r(k_1+2), z_i(k_1+2)], \ldots, [z_r(k_2), z_i(k_2)] \tag{5.33}$$

• SEI:

$$[z_r(k_2+1), z_i(k_2+1)], [z_r(k_2+2), z_i(k_2+2)], \ldots, [z_r(k_3), z_i(k_3)] \tag{5.34}$$

• RL:

$$[z_r(k_3+1), z_i(k_3+1)], [z_r(k_3+2), z_i(k_3+2)], \ldots, [z_r(k_4), z_i(k_4)] \tag{5.35}$$

We will use the above notations to discuss an improved version of the approximate parameter estimation approach discussed in Section 5.5.

5.7.1 Estimation of the Warburg Coefficient

Several frequencies in the Diffusion Arc can be selected to write

$$
\begin{aligned}
z_r(0) - z_r(k_1) &= \sigma \left(\frac{1}{\sqrt{\omega_0}} - \frac{1}{\sqrt{\omega_{k_1}}} \right) \\
z_r(1) - z_r(k_1-1) &= \sigma \left(\frac{1}{\sqrt{\omega_{k_1}}} - \frac{1}{\sqrt{\omega_{k_1-1}}} \right) \\
&\;\;\vdots \\
z_r(n) - z_r(k_1-n) &= \sigma \left(\frac{1}{\sqrt{\omega_n}} - \frac{1}{\sqrt{\omega_{k_1-n}}} \right)
\end{aligned}
\tag{5.36}
$$

where $n < k_1/2$. The observations in (5.36) were selected in such a way that the quantity $z_r(i) - z_r(j)$ could be as high as possible; this strategy is designed to reduce the effect of noise in the observations. The observations (5.36) can be rewritten in matrix form as

$$\tilde{\mathbf{z}} = \mathbf{b}\sigma \tag{5.37}$$

where

$$\tilde{\mathbf{z}} = \begin{bmatrix} z_r(0) - z_r(k_1) \\ z_r(1) - z_r(k_1 - 1) \\ \vdots \\ z_r(n) - z_r(k_1 - n) \end{bmatrix}, \quad \mathbf{b} = \begin{bmatrix} \left(\frac{1}{\sqrt{\omega_0}} - \frac{1}{\sqrt{\omega_{k_1}}} \right) \\ \left(\frac{1}{\sqrt{\omega_{k_1}}} - \frac{1}{\sqrt{\omega_{k_1 - 1}}} \right) \\ \vdots \\ \left(\frac{1}{\sqrt{\omega_n}} - \frac{1}{\sqrt{\omega_{k_1 - n}}} \right) \end{bmatrix} \tag{5.38}$$

The LS estimate of σ is

$$\hat{\sigma} = \frac{(\mathbf{b}^T \tilde{\mathbf{z}})}{(\mathbf{b}^T \mathbf{b})} \tag{5.39}$$

5.7.2 Estimation of the CT Components

Let us denote an impedance measurement in the CT Arc as

$$\begin{aligned} z_r &\triangleq z_r(k) \quad \text{s.t. } k_1 < k \le k_2 \\ z_i &\triangleq z_i(k) \quad \text{s.t. } k_1 < k \le k_2 \end{aligned} \tag{5.40}$$

The measurements in (5.40) will satisfy the following circular equation

$$\begin{aligned} z_r^2 + z_i^2 + a z_r + b &= 0 \\ \left(z_r + \frac{a}{2} \right)^2 + z_i^2 &= \frac{a^2}{4} - b \end{aligned} \tag{5.41}$$

where it was assumed that the center of the circle lies on the real axis (see Figure 5.1). The center of the circle (5.41) is denoted as $(x_{\mathrm{CT}}, 0)$ where

$$x_{\mathrm{CT}} = -\frac{a}{2} \tag{5.42}$$

and the radius of the circle (5.41) is

$$r_{\mathrm{CT}} = \sqrt{\frac{a^2}{4} - b} \tag{5.43}$$

Frequency-Domain Approaches to Battery ECM Identification

Based on (5.41), one can notice that the argument of the square root is always positive. It is now easy to see that the estimate of the resistance R_{CT} is

$$\hat{R}_{\text{CT}} = 2\sqrt{\frac{\hat{a}^2}{4} - \hat{b}} \tag{5.44}$$

where $\hat{a}$ and $\hat{b}$ are estimates of a and b, respectively.

In order to estimate a and b, the pairs of impedance measurements shown in (5.33) can be substituted in (5.41) to get the following sets of equations

$$
\begin{aligned}
-(z_r(k_1 + 1)^2 + z_i(k_1 + 1)^2) &= az_r(k_1 + 1) + b \\
-(z_r(k_1 + 2)^2 + z_i(k_1 + 2)^2) &= az_r(k_1 + 2) + b \\
&\vdots \\
-(z_r(k_2)^2 + z_i(k_2)^2) &= az_r(k_2) + b
\end{aligned}
\tag{5.45}
$$

the above can be written in matrix form as

$$\mathbf{z} = \mathbf{B}\mathbf{x}_{\text{CT}} \tag{5.46}$$

where

$$
\mathbf{z} = \begin{bmatrix} -(z_r(k_1 + 1)^2 + z_i(k_1 + 1)^2) \\ -(z_r(k_1 + 2)^2 + z_i(k_1 + 2)^2) \\ \vdots \\ -(z_r(k_2)^2 + z_i(k_2)^2) \end{bmatrix}, \quad
\mathbf{B} = \begin{bmatrix} z_r(k_1 + 1) & 1 \\ z_r(k_1 + 2) & 1 \\ \vdots \\ z_r(k_2) & 1 \end{bmatrix}, \quad
\mathbf{x}_{\text{CT}} = \begin{bmatrix} a \\ b \end{bmatrix} \tag{5.47}
$$

The least-square estimate of $\mathbf{x}_{\text{CT}}$ is

$$\hat{\mathbf{x}}_{\text{CT}} = \left(\mathbf{B}^T\mathbf{B}\right)^{-1}\mathbf{B}^T\mathbf{z} \tag{5.48}$$

and the estimates of a and b are

$$\hat{a} = \hat{\mathbf{x}}_{\text{CT}}(1), \quad \hat{b} = \hat{\mathbf{x}}_{\text{CT}}(2) \tag{5.49}$$

which will be substituted in (5.44) to estimate R_{CT}.

To estimate C_{DL}, let us start with the total impedance equation

$$Z(\omega) \triangleq j\omega L + R_\Omega + \frac{R_{\text{SEI}}}{1 + j\omega R_{\text{SEI}} C_{\text{SEI}}} + \frac{R_{\text{CT}} + Z_w(j\omega)}{1 + j\omega\left(R_{\text{CT}} + Z_w(j\omega)\right)C_{\text{DL}}} \tag{5.50}$$

At low frequencies, effect of L can be neglected. Let's write the total impedance as real and imaginary parts

$$Z(\omega) \triangleq R_\Omega + \frac{R_{\text{SEI}}(1 - j\omega R_{\text{SEI}}C_{\text{SEI}})}{(1 + \omega^2 R_{\text{SEI}}^2 C_{\text{SEI}}^2)} + \text{term-3} \tag{5.51}$$

where term-3 is given by

$$\text{term-3} = \frac{R_{\text{CT}} + (1 - j)\frac{\sigma}{\sqrt{w}}}{1 + j\omega\left(R_{\text{CT}} + (1 - j)\frac{\sigma}{\sqrt{w}}\right)C_{\text{DL}}} \tag{5.52}$$

$$= \frac{R_{\text{CT}} + \frac{\sigma}{\sqrt{w}} - j\frac{\sigma}{\sqrt{w}}}{1 + j\omega R_{\text{CT}}C_{\text{DL}} + j\omega\frac{\sigma}{\sqrt{w}}C_{\text{DL}} + \sqrt{w}\sigma C_{\text{DL}}} \tag{5.53}$$

$$= \frac{R_{\text{CT}} + \frac{\sigma}{\sqrt{w}} - j\frac{\sigma}{\sqrt{w}}}{1 + \sqrt{w}\sigma C_{\text{DL}} + j(\omega R_{\text{CT}}C_{\text{DL}} + \sqrt{w}\sigma C_{\text{DL}})} \tag{5.54}$$

$$= \frac{c + jd}{a + jb} \tag{5.55}$$

$$= \frac{(c + jd)(a - jb)}{a^2 + b^2} \tag{5.56}$$

where

$$a = 1 + \sqrt{w}\sigma C_{\text{DL}} \tag{5.57}$$

$$b = \omega R_{\text{CT}}C_{\text{DL}} + \sqrt{w}\sigma C_{\text{DL}} \tag{5.58}$$

$$c = R_{\text{CT}} + \frac{\sigma}{\sqrt{w}} \tag{5.59}$$

$$d = -\frac{\sigma}{\sqrt{w}} \tag{5.60}$$

Real-part of term-3 is

$$\text{term-3-r} = \frac{ac + db}{a^2 + b^2} \tag{5.61}$$

Real part of $Z(\omega)$ is then

$$Z_r(\omega) \triangleq R_\Omega + \frac{R_{\text{SEI}}}{(1 + \omega^2 R_{\text{SEI}}^2 C_{\text{SEI}}^2)} + \text{term-3-r} \tag{5.62}$$

Now, solve the following quadratic equation for C_{DL} using MATLAB:

$$(\mathrm{lhs} - R_{\mathrm{CT}} - y) + (2\,\mathrm{lhs}\,wy)C_{\mathrm{DL}} + (\mathrm{lhs}\,w^2R_{\mathrm{CT}}^2 + 2\mathrm{lhs}\,R_{CT}w^2y + 2\mathrm{lhs}w^2y^2)C_{\mathrm{DL}}^2 = 0$$

where

$$\mathrm{lhs} = Z_r - R_\Omega - \frac{R_{\mathrm{SEI}}}{(1 + w^2R_{\mathrm{SEI}}^2C_{\mathrm{SEI}}^2)} \tag{5.63}$$

$$y = \sigma/\sqrt{(\omega)} \tag{5.64}$$

Each C_{DL} at the CT-arc identified point can be determined using the quadratic equation and then averaged.

5.7.3 Estimation of the SEI Components

Let us denote an impedance measurement in the SEI Arc as

$$\begin{aligned}
y_r &\triangleq z_r(k) \quad \text{s.t. } k_2 < k \le k_3 \\
y_i &\triangleq z_i(k) \quad \text{s.t. } k_2 < k \le k_3
\end{aligned} \tag{5.65}$$

The measurements in (5.65) will satisfy the following circular equation

$$y_r^2 + y_i^2 + cy_r + d = 0 \tag{5.66}$$

where it was assumed that the center of the circle lies on the real axis (see Figure 5.1). The center of the circle (5.41) can be denoted as $(x_{\mathrm{SEI}}, 0)$ where

$$x_{\mathrm{SEI}} = -\frac{c}{2} \tag{5.67}$$

and the radius of the circle (5.41) is

$$r_{\mathrm{SEI}} = \sqrt{\frac{c^2}{4} - d} \tag{5.68}$$

By the same reasoning in Section 5.7.2, the argument of the square root can be shown to be always positive. It now is easy to see that the estimate of the resistance R_{CT} is

$$\hat{R}_{\mathrm{SEI}} = 2\sqrt{\frac{\hat{c}^2}{4} - \hat{d}} \tag{5.69}$$

where $\hat{c}$ and $\hat{d}$ are estimates of c and d, respectively. In order to estimate c and d, the pairs of impedance measurements shown in (5.34) can be substituted in (5.66) to get the following sets of equations

$$
\begin{aligned}
-(y_r(k_2 + 1)^2 + y_i(k_2 + 1)^2) &= cy_r(k_2 + 1) + d \\
-(y_r(k_2 + 2)^2 + y_i(k_2 + 2)^2) &= cy_r(k_2 + 2) + d \\
&\vdots \\
-(y_r(k_3)^2 + y_i(k_3)^2) &= cy_r(k_3) + d
\end{aligned}
\tag{5.70}
$$

The above can be written in matrix form as

$$
\mathbf{y} = \mathbf{A}\mathbf{x}_{\mathrm{SEI}} \tag{5.71}
$$

where

$$
\mathbf{y} = \begin{bmatrix} -(y_r(k_2 + 1)^2 + y_i(k_2 + 1)^2) \\ -(y_r(k_2 + 2)^2 + y_i(k_2 + 2)^2) \\ \vdots \\ -(y_r(k_3)^2 + y_i(k_3)^2) \end{bmatrix}, \quad
\mathbf{A} = \begin{bmatrix} y_r(k_2 + 1) & 1 \\ y_r(k_2 + 2) & 1 \\ \vdots & \\ y_r(k_3) & 1 \end{bmatrix}, \quad
\mathbf{x}_{\mathrm{SEI}} = \begin{bmatrix} c \\ d \end{bmatrix}
\tag{5.72}
$$

The LS estimate of $\mathbf{x}_{\mathrm{SEI}}$ is

$$
\hat{\mathbf{x}}_{\mathrm{SEI}} = \left(\mathbf{A}^T \mathbf{A}\right)^{-1} \mathbf{A}^T \mathbf{y} \tag{5.73}
$$

and the estimates of c and d are

$$
\hat{c} = \hat{\mathbf{x}}_{\mathrm{SEI}}(1), \quad \hat{d} = \hat{\mathbf{x}}_{\mathrm{SEI}}(2) \tag{5.74}
$$

which will be substituted in (5.69) to estimate R_{SEI}.

Consider the total impedance equation

$$
Z(\omega) \triangleq j\omega L + R_{\Omega} + \frac{R_{\mathrm{SEI}}}{1 + j\omega R_{\mathrm{SEI}} C_{\mathrm{SEI}}} + \frac{R_{\mathrm{CT}} + Z_{\mathrm{w}}(j\omega)}{1 + j\omega \left(R_{\mathrm{CT}} + Z_{\mathrm{w}}(j\omega)\right) C_{\mathrm{DL}}} \tag{5.75}
$$

At high frequencies, we see the effects of SEI, R_{Ω} and L. Here, the CT-arc and Warburg impedance have negligible effects, as the fourth term of (5.75) is approximately zero.

Then

$$Z(\omega) \triangleq j\omega L + R_\Omega + \frac{R_{\text{SEI}}}{1 + j\omega R_{\text{SEI}} C_{\text{SEI}}} \tag{5.76}$$

Let's write $Z(\omega)$ in terms of real and imaginary components,

$$Z(\omega) \triangleq j\omega L + R_\Omega + \frac{R_{\text{SEI}}(1 - j\omega R_{\text{SEI}} C_{\text{SEI}})}{(1 + j\omega R_{\text{SEI}} C_{\text{SEI}})(1 - j\omega R_{\text{SEI}} C_{\text{SEI}})} \tag{5.77}$$

$$\triangleq j\omega L + R_\Omega + \frac{R_{\text{SEI}}(1 - j\omega R_{\text{SEI}} C_{\text{SEI}})}{(1 + \omega^2 R_{\text{SEI}}^2 C_{\text{SEI}}^2)} \tag{5.78}$$

$$\triangleq R_\Omega + \frac{R_{\text{SEI}}}{(1 + \omega^2 R_{\text{SEI}}^2 C_{\text{SEI}}^2)} + j\frac{(\omega L - \omega R_{\text{SEI}}^2 C_{\text{SEI}})}{(1 + \omega^2 R_{\text{SEI}}^2 C_{\text{SEI}}^2)} \tag{5.79}$$

The real part of the equation is

$$Z_r(\omega) \triangleq R_\Omega + \frac{R_{\text{SEI}}}{(1 + \omega^2 R_{\text{SEI}}^2 C_{\text{SEI}}^2)} \tag{5.80}$$

Now, C_{SEI} can be estimated as follows

$$\hat{C}_{\text{SEI}} = \sqrt{\frac{\frac{R_{\text{SEI}}}{Z_r(\omega) - R_\Omega} - 1}{\omega^2 R_{\text{SEI}}^2}} \tag{5.81}$$

At each SEI identified point, $\hat{C}_{\text{SEI}}$ can be determined and averaged for final estimate of C_{SEI}.

5.7.4   Estimation of Resistance and Inductance

Finally, for a better estimation of the ohmic impedance and the inductance, the values are averaged over a range of high frequencies ($\omega_{k_3} < \omega \le \omega_{k_4}$).

$$\hat{\hat{R}}_\Omega = \frac{1}{k_4 - k_3} \sum_{k=k_3+1}^{k_4} z_r(k) \tag{5.82}$$

$$\hat{\hat{L}} = \frac{1}{(k_4 - k_3)} \sum_{k=k_3+1}^{k_4} \frac{z_i(k)}{\omega_k} \tag{5.83}$$

in which $z_r(k) = \text{Re}\left(Z(\omega_k)\right)$ and $z_i(k) = \text{Im}\left(Z(\omega_k)\right)$.

5.7.5 Feature Point Extraction

It should be stressed that the LS-based (approximate) parameter estimation algorithm presented in this section needs to know the critical frequency values $\omega_{k_0}, \ldots, \omega_{k_4}$. An approach to estimate ω_{k_1}, ω_{k_2}, and ω_{k_3} is presented in [3]. In this approach, a straight line is fitted to the Diffusion Arc with progressively increasing data starting from the lower frequency. As the Diffusion Arc turns into the CT Arc, the correlation of fitting starts to drop. The critical point k_1 is detected by observing the drop in k_1 relative to a predefined threshold. Similarly, a circular curve is fitted to the data starting from the (detected) critical point k_1 and the correlation coefficient is monitored to detect the critical point k_2 based on a predefined threshold. The critical point k_3 can be detected based on the fact that the imaginary part of the Nyquist curve changes its sign at k_3.

In [3], it was assumed that ω_{k_0} and ω_{k_4} span the entire Nyquist spectrum, that is, ω_{k_0} is assumed to be low enough to be on the straight line section of the Diffusion Arc and ω_{k_4} is assumed to be high enough to be on the straight line portion of the RL Arc. Under these assumptions, the values of ω_{k_0} and ω_{k_4} can be identified to be the lowest and highest frequencies, respectively, in the Nyquist spectrum. In other cases, special care must be taken to ensure that the extracted feature points correspond to the assumed arcs. It is also possible that some of the arc may not be present; for example, the SEI Arc may not appear in new batteries, especially when the state of charge is sufficiently high.

5.8 DEMONSTRATION

In this section, computational demonstrations of the proposed ECM parameter identification techniques are presented using simulated and real-world data.

5.8.1 Demonstration Using Simulated Data

In order to obtain the impedance spectrum, simulation is done for a 1000-mAh lithium-ion battery, the parameters of which are taken from a real battery through experimentation. For simulation, a 200 mA (C/5 Rate) DC charging current and an AC perturbation current with a peak of 70 mA are used. A series of 901 single-sine waves was used with different frequencies in the range of 0.01 Hz to 10 kHz ($k = 901$ frequencies) for better resolution and a more precise demonstration of an industrial Li-ion battery. The battery charging current, $I_{\rm B}(t)$, is represented as

$$I_{\rm B}(t) = \begin{cases} I_{\rm dc} + I_{\rm m}sin(\omega_1 t), & t < t_1 \\ I_{\rm dc} + I_{\rm m}sin(\omega_2 t), & t_1 \leq t < t_2 \\ \vdots \\ I_{\rm dc} + I_{\rm m}sin(\omega_k t), & t_{k-1} \leq t < t_k \end{cases} \tag{5.84}$$

where $I_{\rm dc}$ is the DC, $I_{\rm m}$ is the peak of perturbation current, and each t_k is selected such that $t_k - t_{k-1} = \frac{1}{f_k}$ (i.e., each frequency is made to have one full cycle of data).

For measuring the noise effects, it is assumed that the measured current and voltage are corrupted with Gaussian noise of the same noise variances, that is,

$$\sigma_v = \sigma_i = I_m 10^{\left(-\frac{\text{SNR}}{20}\right)} \tag{5.85}$$

where SNR varies from 0 to 50 dB (0, 10, 20, 30, 40, 50 dB).

The impedance spectrum is derived for each case (each level of noise) and algorithms explained in Sections 5.4, 5.5, and 5.7 are applied to estimate ECM parameters. The performance of each algorithm is quantified in terms of the normalized percentage mean square error, simply referred to hereafter as Error%. For example, the *Error (%)* of estimating the Ohmic resistance by the proposed approach is defined as

$$\text{Error}(\%) = \frac{\left| R_\Omega - \hat{R}_\Omega \right|}{R_\Omega} 100 \tag{5.86}$$

Each reported error measure is averaged over 100 Monte Carlo runs.

Figures 5.4 through 5.7 present the percentage error for each of the approaches discussed in this chapter. Figure 5.4 compares the performance in estimating the Warburg coefficient (σ). An explanation of the performance loss by the approximate approach (Section 5.5) could be that it used only two points to find the slope of a line to estimate σ. However, the improved approach (Section 5.7) used many pairs of points and resulted in better estimation error. The reason for the failure of the nonlinear LS approach could be attributable to the severe nonlinearity in the model when it comes to estimating the Warburg coefficient.

Figures 5.5(a) and 5.5(b) show the performance comparison of the three approaches presented in this chapter for the $R_{\rm CT}$ and $C_{\rm DL}$ estimation.

Figures 5.6(a) and 5.6(b) show a similar comparison for the $R_{\rm SEI}$ and $C_{\rm SEI}$ estimation. These two figures exhibit the performance trade-off of the different approaches presented in this chapter and highlight the need to develop robust approaches to estimate ECM parameters.

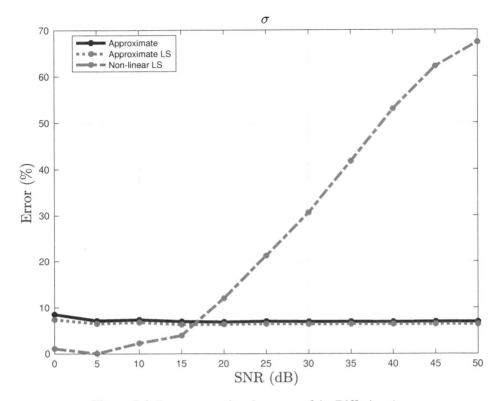

Figure 5.4 Parameter estimation error of the Diffusion Arc.

Figure 5.7(a) shows the estimation errors corresponding to the Ohmic resistance, R_Ω. All the algorithms have excellent performance (error rate lower than 1%) in estimating R_Ω. This is due to the linear relationship of R_Ω to the measurements in (5.2). Figure 5.7(b) shows the comparison of different estimators in the estimation of stray inductance L.

For a comprehensive analysis of the performance of each three approaches presented in this chapter, the extracted parameters from each method are used to generate the impedance spectrum. Figure 5.8(a) compares the performance of the previous approach and the proposed algorithm in estimating the impedance spectrum at SNR = 30 dB. Figure 5.8(b) shows the performance comparison of all three approaches in a severe but practical case (low SNR). In the presence of a high level of noise, the approximate LS

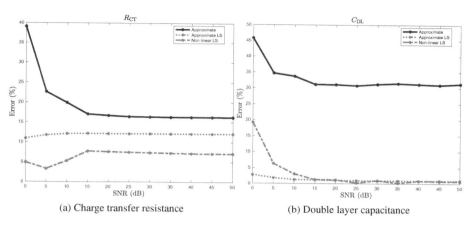

(a) Charge transfer resistance

(b) Double layer capacitance

Figure 5.5 (a, b) Parameter estimation of the CT Arc.

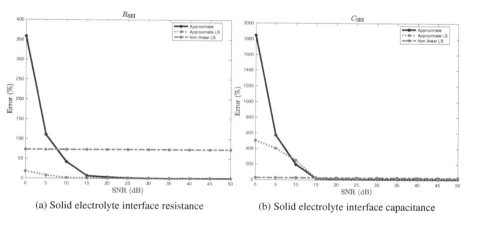

(a) Solid electrolyte interface resistance

(b) Solid electrolyte interface capacitance

Figure 5.6 (a, b) Parameter estimation of the SEI Arc.

approach is observed to outperform both the nonlinear LS approach and the approximate approach, as shown in Figure 5.8(b).

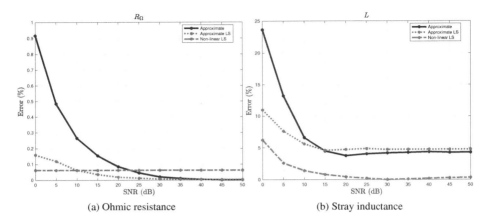

(a) Ohmic resistance   (b) Stray inductance

Figure 5.7   (a, b) Parameter estimation of the RL Arc.

5.8.2   Demonstration Using Real Data

In this section, ECM parameter extraction is demonstrated using a real-world experiment. For this, an Arbin battery cycler with Gamry interface 5000P EIS device was used. Figure 5.9 shows these devices in an experimental setup; the computer screen shows the interface of the Arbin software that allows to the collection of impedance data from the Gamry device. A relatively new cylindrical Li-ion battery (LG INR18650 MJ1) is used in the experiment. The specifications of this battery are shown in Table 5.1. The state of charge of the battery was nearly empty when the experiment was performed.

First, the Gamry EIS device was programmed to use zero DC with a 50 mA sinusoid superimposed on it. In a second experiment, right after the first experiment, the experiment was repeated with a 200 mA charging current and a 50 mA sinusoid superimposed on it. The output of the Gamry EIS device is the real and imaginary values of the measured impedance, as shown in Figure 5.10. The parameter estimation algorithms proposed in Sections 5.4 and 5.7 were applied to the impedance measurement to estimate the battery's AR-ECM parameters. Table 5.2 shows the estimated values of the AR-ECM parameters based on the nonlinear LS approach. It can be seen in Table 5.2 that the estimated parameters are slightly different when there is a DC compared to the zero-mean current experiment. It is assumed in this chapter that the DC values of both voltage and current are zero. However, in reality, the nonzero DC is more practical (that can be superimposed during battery charging) and this results in performance loss. A

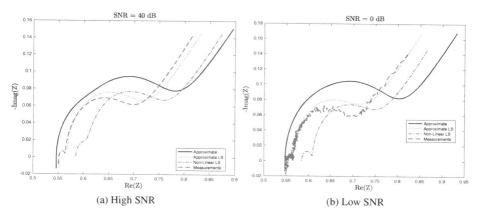

Figure 5.8 (a, b) Performance comparison where the estimated parameters were used to reconstruct impedance spectrum.

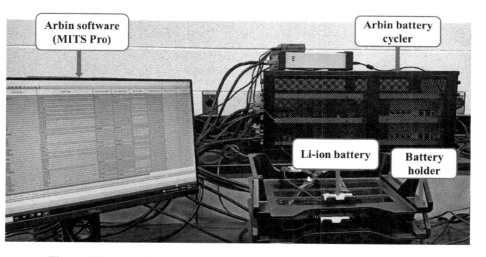

Figure 5.9 Experimental setup. Reproduced with permission from [4].

detailed analysis of such performance losses is not discussed in this chapter. Instead, the experiment presented in this section is intended to make one aware of the discrepancies of estimated parameters in the presence of nonzero DC. Table 5.3 summarizes the results of ECM parameter estimation using the approximation LS approach. It can be noticed that the estimated Warburg coefficient is significantly different in Tables 5.2 and 5.3. In the absence of true values, according to the simulation analysis presented in Section 5.8.1, Table 5.3 can be considered to provide the most approximate ECM parameters of the battery.

Table 5.1

Battery Specifications

Specification	Value (Unit)
Nominal capacity	3500 mAh
Maximum current	10A
Nominal voltage (V_{nom})	3.7V
Height	65 mm
Diameter	18 mm
Weight	46.5g

Table 5.2

Estimated Parameters (Nonlinear LS)

Parameter	Value (0 DC)	Value (-0.2 DC)
R_Ω	412.5 mΩ	411.3 mΩ
R_{CT}	2.6848 mΩ	3.22 mΩ
C_{DL}	1.000F	1.000F
L	1.1588×10^{-06} H	1.1534×10^{-06} H
σ	0.0584	0.0656

5.9 SUMMARY

This chapter introduced frequency-domain approaches to battery equivalent circuit model parameter estimation using EIS. In EIS, an excitation signal (either voltage or current) is applied to the battery and its response (current or voltage) is measured. This procedure is repeated and the amplitude and phase of the frequency response are computed at various (fixed) frequencies spanning very low frequencies in fractions

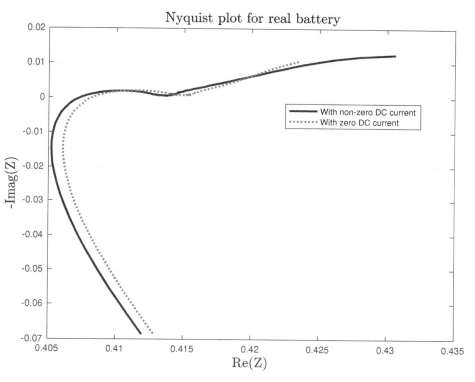

Figure 5.10 Nyquist plot for the real battery. The Nyquist plot shown in the solid line depicts the impedance response of the battery with 0.2A DC. The impedance response of the battery with 0A DC is shown by the dotted line.

of Hz and very high frequencies in several MHz. Based on the obtained responses at wide-ranging frequencies, the ECM parameters can be estimated. This chapter outlined three different approaches to estimating the ECM parameters based on the frequency response. The first approach was based on the nonlinear least squares estimation which requires significant computational resources. Also, the nonlinear least squares approach was shown to be susceptible to local convergence even at high SNR regions. It was also shown that distinct portions of the impedance spectrum can be extracted to estimate ECM parameters.

Table 5.3
Estimated Parameters (Approximate LS)

Parameter	Value (0 DC)	Value (-0.2 DC)
R_Ω	410.7 mΩ	409.7 mΩ
R_{CT}	3.1 mΩ	3.3 mΩ
C_{DL}	1.0278F	0.8740F
L	1.1142×10^{-06} H	1.1107×10^{-06} H
σ	0.0020	0.0041

The EIS approach to battery analysis is a time-consuming process. Measurements at low frequencies especially incur a significant delay. Significant recent research work is focused on reducing the experimental time of the EIS approach for battery analysis.

5.10   BIBLIOGRAPHICAL NOTES

A well-detailed analysis of EIS with numerical examples can be found in [2]. Free software for impedance spectrum analysis can be downloaded from [5]; this approach is based on the classic work [6].

References

[1] E. Barsoukov, and J. R. Macdonald, *Impedance Spectroscopy: Theory, Experiment, and Applications* (2nd. ed.), John Wiley & Sons, Inc., New York, 2008.

[2] M. E. Orazem, and B. Tribollet, *Electrochemical Impedance Spectroscopy,* John Wiley & Sons, Inc., New York, 2008.

[3] M. Abaspour, K. R. Pattipati, B. Shahrrava, and B. Balasingam, "Robust Approach to Battery Equivalent-Circuit-Model Parameter Extraction Using Electrochemical Impedance Spectroscopy," *Energies,* Vol. 15, No. 23, pp. 9251, 2022.

[4] P. Kumar, C. Fuerth, G. Rankin, K. R. Pattipati, and B. Balasingam, "Hardware in the Loop Demonstration of Battery Surface Temperature Prediction," *2022 IEEE International Conference on Environment and Electrical Engineering and 2022 IEEE Industrial and Commercial Power Systems Europe (EEEIC/I&CPS Europe),* pp. 1-6, 2022.

[5] Impedance Spectrum Analysis Software, https://jrossmacdonald.com/levmlevmw/ (accessed Dec. 2021).

[6] J. R. Macdonald, and J. A. Garber, "Analysis of impedance and admittance data for solids and liquids," *Journal of the Electrochemical Society,* Vol. 124, No. 7, pp. 1022, 1977.

Chapter 6

Time-Domain Approaches to Battery ECM Identification

6.1 INTRODUCTION

It was described in Chapter 3 that the electrical system response of a battery can be represented through an electrical ECM consisting of resistors, capacitors, and inductors. It is also found through experimental studies that the electrical response of a battery varies with temperature, SOC, and age, implying the variation of ECM parameters. Due to this, prior estimation or characterization of ECM parameters has little use in battery management systems. The ECM parameters need to be continuously estimated for effective battery management.

Several aspects of battery management require the ECM parameters of a battery. For SOC estimation, the ECM parameters are used to model the voltage drop within the battery. The ECM parameters are needed to compute the remaining mileage of an electric vehicle. During battery charging, the ECM parameters are needed to determine the charging current for accurate and fast charging. In battery thermal management, the ECM parameters are used to compute the heat generated within the battery so that the surface temperature can be predicted.

This chapter presents a time-domain approach to ECM parameter estimation. This approach is comparable to the frequency-domain approach to ECM parameter estimation presented in Chapter 5. Both of these approaches are designed in a way that they are independent of the SOC of the battery. That is, both of the ECM parameter estimation approaches presented in this book can be applied to estimate ECM parameters without the knowledge of the battery SOC. This chapter also highlights the need for the excitation signal design to improve the accuracy of ECM parameter estimation. It is shown that a theoretical bound on the estimation error can be derived for a given excitation signal. Based on this bound, the parameters of the excitation signal can be selected to improve the estimation accuracy.

The remainder of the chapter is structured as follows. In Section 6.2, the mathematical derivation of the measurement model that is based only on the measured voltage and current through the battery state is presented. In Section 6.3, the derivation of the measurement model is extended to different model orders. Section 6.4 describes the proposed parameter estimation method and Section 6.5 contains the theoretical performance analysis of the proposed method. Section 6.6 summarizes the results of the testing approaches for simulated and real data.

6.2   SIGNAL MODEL OF A BATTERY

Figure 6.1 shows four different approximations of an ECM to be considered in this chapter. The derivations presented in this section are based on the most general model shown in Figure 6.1(d). Section 6.3 shows how these derivations can be applied to the other three models.

The measured current through the battery is written as

$$z_i[k] = i[k] + n_i[k] \tag{6.1}$$

where $i[k]$ is the true current through the battery and $n_i[k]$ is the current measurement noise, which is assumed to be zero means and has a standard deviation (s.d.) σ_i. The measured voltage across the battery is

$$z_v[k] = v[k] + n_v[k] \tag{6.2}$$

where $v[k]$ is the true voltage across the battery and $n_v[k]$ is the voltage measurement noise, which is assumed to be zero mean with s.d. σ_v.

For the ECM model in Figure 6.1(d), the true voltage across the battery, $v[k]$, is written as the sum of the voltage drop across the internal components, R_0, R_1, R_2, and the EMF, $v_o[k]$. Hence, (6.2), can be rewritten as

$$z_v[k] = i[k]R_0 + x_{i_1}[k]R_1 + x_{i_2}[k]R_2 + v_o[k] + n_v[k] \tag{6.3}$$

where the currents through the resistors R_1 and R_2 can be written in the following form

$$x_{i_1}[k+1] \triangleq i_1[k+1] = \alpha_1 i_1[k] + (1 - \alpha_1)i[k] \tag{6.4}$$

$$x_{i_2}[k+1] \triangleq i_2[k+1] = \alpha_2 i_2[k] + (1 - \alpha_2)i[k] \tag{6.5}$$

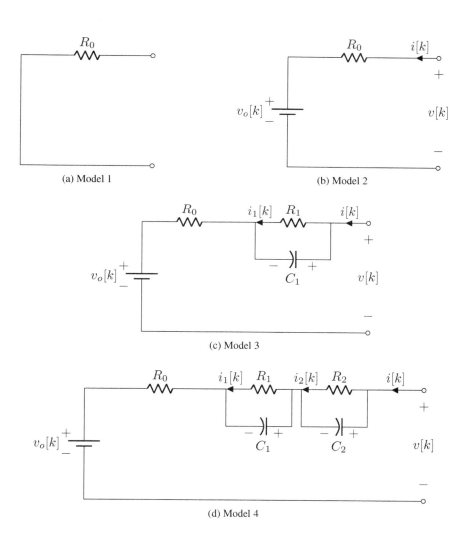

(a) Model 1

(b) Model 2

(c) Model 3

(d) Model 4

Figure 6.1 Different ECM model orders.

where

$$\alpha_1 \triangleq e^{-\frac{\Delta}{R_1 C_1}} \tag{6.6}$$

$$\alpha_2 \triangleq e^{-\frac{\Delta}{R_2 C_2}} \tag{6.7}$$

and Δ is the sampling interval. By substituting the measured current $z_i[k]$ for $i[k]$, the currents in (6.4) and (6.5) can be rewritten as follows

$$x_{i_1}[k+1] = \alpha_1 x_{i_1}[k] + (1-\alpha_1)z_i[k] - (1-\alpha_1)n_i[k] \tag{6.8}$$

$$x_{i_2}[k+1] = \alpha_2 x_{i_2}[k] + (1-\alpha_2)z_i[k] - (1-\alpha_2)n_i[k] \tag{6.9}$$

Now, using (6.1), (6.3) can be rewritten in the z-domain as follows

$$Z_v[z] = Z_i[z]R_0 + X_{i_1}[z]R_1 + X_{i_2}[z]R_2 + V_o[z] + N_v[z] - R_0 N_i[z] \tag{6.10}$$

Next, let us rewrite (6.8) in the z-domain

$$zX_{i_1}[z] = \alpha_1 X_{i_1}[z] + (1-\alpha_1)Z_i[z] - (1-\alpha_1)N_i[z] \tag{6.11}$$

which yields

$$X_{i_1}[z] = \frac{1-\alpha_1}{z-\alpha_1}\left(Z_i[z] - N_i[z]\right) \tag{6.12}$$

and, similarly for (6.9),

$$X_{i_2}[z] = \frac{1-\alpha_2}{z-\alpha_2}\left(Z_i[z] - N_i[z]\right) \tag{6.13}$$

By substituting (6.12) and (6.13) into (6.10), one gets

$$Z_v[z] = Z_i[z]R_0 + \frac{1-\alpha_1}{z-\alpha_1}Z_i[z]R_1 + \frac{1-\alpha_2}{z-\alpha_2}Z_i[z]R_2 + V_o[z]$$
$$+ N_v[z] - \left(R_0 + \frac{1-\alpha_1}{z-\alpha_1}R_1 + \frac{1-\alpha_2}{z-\alpha_2}R_2\right)N_i[z] \tag{6.14}$$

Rearranging (6.14) and converting it back to the time domain, we get

$$z_v[k] = \alpha z_v[k-1] - \boldsymbol{\eta} z_v[k-2] + R_0 z_i[k] - \check{R}_1 z_i[k-1] + \check{R}_2 z_i[k-2]$$
$$+ V_o[k] + \bar{n}_i[k] + \bar{n}_v[k] \tag{6.15}$$

where

$$\alpha = \alpha_1 + \alpha_2$$
$$\boldsymbol{\eta} = \alpha_1 \alpha_2$$
$$\check{R}_1 = (\alpha_1 + \alpha_2)R_0 - (1 - \alpha_1)R_1 - (1 - \alpha_2)R_2$$
$$\check{R}_2 = \alpha_1 \alpha_2 R_0 - \alpha_2(1 - \alpha_1)R_1 - \alpha_1(1 - \alpha_2)R_2$$
$$V_o[k] = v_o[k] - \alpha v_o[k-1] + \boldsymbol{\eta} v_o[k-2]$$
$$\bar{n}_v[k] = n_v[k] - \alpha n_v[k-1] + \boldsymbol{\eta} n_v[k-2]$$
$$\bar{n}_i[k] = -R_0 n_i[k] + \check{R}_1 n_i[k-1] - \check{R}_2 n_i[k-2]$$

Consider $V_o[k]$ is constant over a small window of time k, and then $V_o[k] \approx V_o$. Therefore, (6.15) can be rewritten as

$$z_v[k] = \alpha z_v[k-1] - \boldsymbol{\eta} z_v[k-2] + R_0 z_i[k] - \check{R}_1 z_i[k-1] + \check{R}_2 z_i[k-2]$$
$$+ V_o + \bar{n}_i[k] + \bar{n}_v[k] \qquad (6.16)$$

Now let us rewrite (6.16) in the following form

$$z_v[k] = \mathbf{a}[k]^T \mathbf{b} + n_D[k] \qquad (6.17)$$

where the observation model $\mathbf{a}[k]^T$ and the model parameter vector $\mathbf{b}$ for the ECM model are given by

$$\mathbf{a}[k]^T = \mathbf{a}_4[k]^T$$
$$\triangleq \begin{bmatrix} z_v[k-1] & -z_v[k-2] & z_i[k] & -z_i[k-1] & z_i[k-2] & 1 \end{bmatrix} \qquad (6.18)$$
$$\mathbf{b} = \mathbf{b}_4$$
$$\triangleq [\alpha \quad \boldsymbol{\eta} \quad R_0 \quad \check{R}_1 \quad \check{R}_2 \quad V_o]^T \qquad (6.19)$$

The subscripts 4 in (6.18) and (6.19) indicate that the model corresponds to Model 4, as in Figure 6.1(d). ECMs 1–3 are discussed later in Section 6.3. The noise in the voltage drop in (6.17) is written as

$$n_D[k] \triangleq \bar{n}_i[k] + \bar{n}_v[k] \qquad (6.20)$$

which has the following autocorrelation

$$R_{n_D}(l) = E\left(n_D[k]n_D[k-l]\right), \quad l = 0, 1, 2, \ldots \qquad (6.21)$$

where l denotes the sampling delay; for $l = 0$, the resulting correlation is the autocorrelation of the noise without time delay. For $l \neq 0$, the nonzero value of the autocorrelation indicates time-correlated noise in the observation model (6.17). For the Kalman filter, the noise must not have a time correlation; when there is time correlation, special procedures must be followed to account for such correlation in the model [1].

This autocorrelation $R_{n_D}(l)$ for different values of l are given below:

$$
\begin{aligned}
R_{n_D}(0) &= E\Big\{ n_D[k]n_D[k] \Big\} \\
&= E\Big\{ \big\{ n_v^2[k] + \alpha^2 n_v^2[k-1] + \eta^2 n_v^2[k-2] \\
&\qquad + R_0^2 n_i^2[k] + \check{R}_1^2 n_i^2[k-1] + \check{R}_2^2 n_i^2[k-2] \big\} \Big\} \\
&= (1 + \alpha^2 + \eta^2)\sigma_v^2 + (R_0^2 + \check{R}_1^2 + \check{R}_2^2)\sigma_i^2
\end{aligned}
\tag{6.22}
$$

$$
\begin{aligned}
R_{n_D}(1) &= E\Big\{ n_D[k]n_D[k-1] \Big\} \\
&= E\Big\{ \big(-\alpha n_v[k-1] + \eta n_v[k-2] + \check{R}_1 n_i[k-1] - \check{R}_2 n_i[k-2] \big) \\
&\qquad \big(n_v[k-1] - \alpha n_v[k-2] - R_0 n_i[k-1] + \check{R}_1 n_i[k-2] \big) \Big\} \\
&= -\alpha(1 + \eta)\sigma_v^2 - \check{R}_1(R_0 + \check{R}_2)\sigma_i^2
\end{aligned}
\tag{6.23}
$$

$$
\begin{aligned}
R_{n_D}(2) &= E\Big\{ n_D[k]n_D[k-2] \Big\} \\
&= E\Big\{ \big(\eta n_v[k-2] - \check{R}_2 n_i[k-2] \big)\big(n_v[k-2] - R_0 n_i[k-2] \big) \Big\} \\
&= \eta\sigma_v^2 + R_0\check{R}_2\sigma_i^2
\end{aligned}
\tag{6.24}
$$

All the noise autocorrelation values can be summarized as follows:

$$
R_{n_D}(l) \triangleq E\Big\{ n_D[k]n_D[k-l] \Big\}
$$
$$
\begin{cases}
(1 + \alpha^2 + \eta^2)\sigma_v^2 + (R_0^2 + \check{R}_1^2 + \check{R}_2^2)\sigma_i^2 & |l| = 0 \\
-\alpha(1 + \eta)\sigma_v^2 - \check{R}_1(R_0 + \check{R}_2)\sigma_i^2 & |l| = 1 \\
\eta\sigma_v^2 + R_0\check{R}_2\sigma_i^2 & |l| = 2 \\
0 & |l| > 2
\end{cases}
\tag{6.25}
$$

6.3 ECM IDENTIFICATION OF DIFFERENT MODEL ORDERS

The four models are explained below:

- Model 1: A series resistance only (Figure 6.1(a)).
- Model 2: A series resistance and the battery (Figure 6.1(b)).
- Model 3: A series resistance, the battery, and a single RC circuit (Figure 6.1(c)).
- Model 4: A series resistance, the battery, and two RC circuits (Figure 6.1(d)).

The measured voltage of each of the four equivalent circuit models shown in Figure 6.1 can be written in the following form:

$$z_v[k] = \mathbf{a}[k]^T \mathbf{b} + n_D[k] \tag{6.26}$$

where

$$\mathbf{a}[k]^T = \begin{cases} \mathbf{a}_1^T[k] & \text{Model 1} \\ \mathbf{a}_2^T[k] & \text{Model 2} \\ \mathbf{a}_3^T[k] & \text{Model 3} \\ \mathbf{a}_4^T[k] & \text{Model 4} \end{cases} \qquad \mathbf{b} = \begin{cases} \mathbf{b}_1 & \text{Model 1} \\ \mathbf{b}_2 & \text{Model 2} \\ \mathbf{b}_3 & \text{Model 3} \\ \mathbf{b}_4 & \text{Model 4} \end{cases} \tag{6.27}$$

where

$$\mathbf{a}_1^T[k] = z_i[k] \qquad \mathbf{a}_2^T[k] = [z_i[k] \ \ 1]$$

$$\mathbf{a}_3^T[k] = \left[z_v[k-1] \ \ z_i[k] \ \ -z_i[k-1] \ \ 1 \right]$$

$$\mathbf{a}_4^T[k] = \left[z_v[k-1] \ \ -z_v[k-2] \ \ z_i[k] \ \ -z_i[k-1] \ \ z_i[k-2] \ \ 1 \right]$$

$$\mathbf{b}_1 = R_0 \qquad\qquad\qquad \mathbf{b}_2 = [R_0 \ \ V_o]^T$$

$$\mathbf{b}_3 = \left[\alpha_1 \ R_0 \ \check{R}_1 \ V_o \right]^T \qquad \mathbf{b}_4 = \left[\alpha \ \eta \ R_0 \ \check{R}_1 \ \check{R}_2 \ V_o \right]^T$$

For each of the above model complexities, the noise term $n_D[k]$ is expressed in terms of $\bar{n}_i[k]$ and $\bar{n}_v[k]$ as follows:

$$\bar{n}_v[k] = \begin{cases} \bar{n}_{v1}[k] & \text{Model 1} \\ \bar{n}_{v2}[k] & \text{Model 2} \\ \bar{n}_{v3}[k] & \text{Model 3} \\ \bar{n}_{v4}[k] & \text{Model 4} \end{cases} \qquad \bar{n}_i[k] = \begin{cases} \bar{n}_{i1}[k] & \text{Model 1} \\ \bar{n}_{i2}[k] & \text{Model 2} \\ \bar{n}_{i3}[k] & \text{Model 3} \\ \bar{n}_{i4}[k] & \text{Model 4} \end{cases} \tag{6.28}$$

where

$$\bar{n}_{v1}[k] = \bar{n}_{v2}[k] = n_v[k]$$
$$\bar{n}_{v3}[k] = n_v[k] - \alpha_1 n_v[k-1]$$
$$\bar{n}_{v4}[k] = n_v[k] - (\alpha_1 + \alpha_2)n_v[k-1] + \alpha_1\alpha_2 n_v[k-2]$$

$$\bar{n}_{i1}[k] = \bar{n}_{i2}[k] = -R_0 n_i[k]$$
$$\bar{n}_{i3}[k] = -R_0 n_i[k] + \check{R}_1 n_i[k-1]$$
$$\bar{n}_{i4}[k] = -R_0 n_i[k] + \check{R}_1 n_i[k-1] - \check{R}_2 n_i[k-2]$$

The autocorrelation for all four model orders is as follows:

$$R_{n_D}(0) = E\left\{n_D[k]n_D[k]\right\}$$
$$= \begin{cases} \sigma_v^2 + R_0^2\sigma_i^2 & \text{Model 1} \\ \sigma_v^2 + R_0^2\sigma_i^2 & \text{Model 2} \\ (1+\alpha_1^2)\sigma_v^2 + \left(R_0^2 + \check{R}_1^2\right)\sigma_i^2 & \text{Model 3} \\ (1+\alpha^2+\eta^2)\sigma_v^2 + \left(R_0^2 + \check{R}_1^2 + \check{R}_2^2\right)\sigma_i^2 & \text{Model 4} \end{cases} \quad (6.29)$$

$$R_{n_D}(1) = E\left\{n_D[k]n_D[k-1]\right\}$$
$$= \begin{cases} 0 & \text{Model 1} \\ 0 & \text{Model 2} \\ -\alpha_1\sigma_v^2 - R_0\check{R}_1\sigma_i^2 & \text{Model 3} \\ -\alpha(1+\eta)\sigma_v^2 - \check{R}_1(R_0 + \check{R}_2)\sigma_i^2 & \text{Model 4} \end{cases} \quad (6.30)$$

$$R_{n_D}(2) = E\left\{n_D[k]n_D[k-2]\right\} = \begin{cases} 0 & \text{Model 1} \\ 0 & \text{Model 2} \\ 0 & \text{Model 3} \\ \eta\sigma_v^2 + R_0\check{R}_2\sigma_i^2 & \text{Model 4} \end{cases} \quad (6.31)$$

The autocorrelation $R_{n_D}(l)$ is zero when $l > 2$ for all four models.

6.4 PARAMETER ESTIMATION METHOD

The measurements are grouped into batches of equal length L_b. Using (6.26), the vector observation model is rewritten for a particular batch of data of length L_b.

$$\mathbf{z}_v[\kappa] = \mathbf{H}[\kappa]^T \mathbf{b} + \mathbf{n}_D[\kappa] \tag{6.32}$$

where κ denotes the batch number,

$$\mathbf{z}_v[\kappa] = \begin{bmatrix} z_v[(\kappa-1)L_b+1] \\ z_v[(\kappa-1)L_b+2] \\ \vdots \\ z_v[\kappa L_b] \end{bmatrix}, \quad \mathbf{n}_D[\kappa] = \begin{bmatrix} n_D[(\kappa-1)L_b+1] \\ n_D[(\kappa-1)L_b+2] \\ \vdots \\ n_D[\kappa L_b] \end{bmatrix}$$

$$\mathbf{H}[\kappa] = \begin{bmatrix} \mathbf{a}[(\kappa-1)L_b+1] \\ \mathbf{a}[(\kappa-1)L_b+2] \\ \vdots \\ \mathbf{a}[\kappa L_b] \end{bmatrix}, \quad \mathbf{a}[k]^T = \begin{cases} \mathbf{a}_1^T[k] & \text{Model 1} \\ \mathbf{a}_2^T[k] & \text{Model 2} \\ \mathbf{a}_3^T[k] & \text{Model 3} \\ \mathbf{a}_4^T[k] & \text{Model 4} \end{cases}$$

The correlation matrix of the noise vector $\mathbf{n}_D[\kappa]$ is written as

$$E\left(\mathbf{n}_D[\kappa]\mathbf{n}_D[\kappa]^T\right) = \mathbf{R}_{\mathbf{n}_D}[\kappa] \tag{6.33}$$

where $\mathbf{R}_{\mathbf{n}_D}[\kappa]$ is a banded symmetric Toeplitz matrix, which is diagonal for Model 1 and Model 2, tridiagonal for Model 3, and pentadiagonal for Model 4. The diagonal entry of $\mathbf{R}_{\mathbf{n}_D}[\kappa]$ is given by $R_{n_D}(1)$, and the first off-diagonal entry is given by $R_{n_D}(1)$, and the second off-diagonal entry is given by $R_{n_D}(2)$ in (6.29) to (6.31). All the other off-diagonal elements of $\mathbf{R}_{\mathbf{n}_D}[\kappa]$ are zero.

Given the κth batch of observations, the LS estimate of $\mathbf{b}$ can be written as

$$\hat{\mathbf{x}}_{\text{LS}}[\kappa] = \left(\mathbf{H}[\kappa]^T \mathbf{R}_{\mathbf{n}_D}[\kappa]^{-1} \mathbf{H}[\kappa]\right)^{-1} \mathbf{H}[\kappa]^T \mathbf{R}_{\mathbf{n}_D}[\kappa]^{-1} \mathbf{z}_v[\kappa] \tag{6.34}$$

It can be shown that the covariance of the LS estimation error is

$$P[\kappa] = \left(\mathbf{H}[\kappa]^T \mathbf{R}_{\mathbf{n}_D}[\kappa]^{-1} \mathbf{H}[\kappa]\right)^{-1} \tag{6.35}$$

When the parameter $\mathbf{b}$ needs to be estimated using more data, the batch length L_b increases, resulting in significantly high computational complexity. Rather than increasing L_b, a recursive least-square (RLS) algorithm can be employed to achieve the same performance without significantly increasing computational load. Algorithm 6.1 summarizes one iteration of the RLS algorithm. The input to this algorithm is the estimate $\hat{\mathbf{x}}[\kappa]$ and the error covariance $\mathbf{P}[\kappa]$ from the prior batch, the new measurement $\mathbf{z}[\kappa+1]$, and the new measurement model $\mathbf{H}[\kappa+1]$. The outputs is the new estimate $\hat{\mathbf{x}}[\kappa+1]$ and the updated error covariance $\mathbf{P}[\kappa+1]$.

Algorithm 6.1 $\left[\hat{\mathbf{x}}[\kappa+1], \mathbf{P}[\kappa+1]\right] = \mathrm{RLS}\left[\hat{\mathbf{x}}[\kappa], \mathbf{P}[\kappa], \mathbf{H}[\kappa+1], \mathbf{z}[\kappa+1]\right]$

1: Update residual covariance:
$\mathbf{S}[\kappa+1] = \mathbf{H}[\kappa+1]\mathbf{P}[\kappa]\mathbf{H}[\kappa+1]^T + \mathbf{R}[\kappa+1]$
2: Update gain:
$\mathbf{W}[\kappa+1] = \mathbf{P}[\kappa]\mathbf{H}[\kappa+1]^T\mathbf{S}[\kappa+1]^{-1}$
3: Update parameter:
$\hat{\mathbf{x}}[\kappa+1] = \hat{\mathbf{x}}[\kappa] + \mathbf{W}[\kappa+1]\left(\mathbf{z}[\kappa+1] - \mathbf{H}[\kappa+1]\hat{\mathbf{x}}[\kappa]\right)$
4: Update information:
$\mathbf{P}^{-1}[\kappa+1] = \mathbf{P}^{-1}[\kappa] + \mathbf{H}[\kappa+1]^T\mathbf{R}[\kappa+1]^{-1}\mathbf{H}[\kappa+1]$

The calculation of the model parameters from the estimates $\hat{\mathbf{x}}_{LS}[\kappa]$ is given below for Model 3. From (6.28), the estimate has four elements:

$$\hat{\mathbf{x}}_{LS}[\kappa] = \mathbf{b}_3 = \begin{bmatrix} \alpha_1 & R_0 & \check{R}_1 & V_o \end{bmatrix}^T$$

After estimation, the parameters of the ECM Model 3, R_1 and C_1, are recovered as follows:

$$R_1 = \frac{(\check{R}_1 - \alpha_1 R_0)}{(\alpha_1 - 1)}, \qquad C_1 = \frac{-\Delta}{R_1 \ln \alpha_1}$$

6.5   PERFORMANCE ANALYSIS

In this section, a theoretical performance analysis of the proposed parameter estimation algorithm is developed. For the analysis in this section, we will assume that the current is perfectly known. For linear observation model (6.32) under Gaussian noise assumption, the Cramer-Rao Lower Bound (CRLB) [1] serves as the lower bound on the estimation error covariance. It can be shown that, for the observation model (6.32), the CRLB is

$$\mathrm{CRLB} = \left(\mathbf{H}[\kappa]^T\boldsymbol{\Sigma}^{-1}\mathbf{H}[\kappa]\right)^{-1} = \sigma_v^2\left(\mathbf{H}[\kappa]^T\mathbf{H}[\kappa]\right)^{-1} \tag{6.36}$$

that is,

$$E\left((\hat{\mathbf{b}} - \mathbf{b})(\hat{\mathbf{b}} - \mathbf{b})^T\right) \geq \text{CRLB} \tag{6.37}$$

where $\hat{\mathbf{b}}$ denotes an estimate of $\mathbf{b}$.

Now, let us focus on the CRLB corresponding to Model 2 in Figure 6.1 for an in-depth analysis. For this model, $\mathbf{H}[\kappa]^T\mathbf{H}[\kappa]$ can be expanded as follows

$$\mathbf{H}[\kappa]^T\mathbf{H}[\kappa] = \begin{bmatrix} \sum_{k=1}^L i(k)^2 & \sum_{k=1}^L i(k) \\ \sum_{k=1}^L i(k) & L \end{bmatrix} \tag{6.38}$$

where it is assumed $\mathbf{z}_i(k) = i(k)$ to simplify the analysis.

Now $\left(\mathbf{H}[\kappa]^T\mathbf{H}[\kappa]\right)^{-1}$ can be simplified as

$$\left(\mathbf{H}[\kappa]^T\mathbf{H}[\kappa]\right)^{-1} = \left(\frac{1}{|\mathbf{H}[\kappa]^T\mathbf{H}[\kappa]|}\right) \begin{bmatrix} L & -\sum_{k=1}^L i(k) \\ -\sum_{k=1}^L i(k) & \sum_{k=1}^L i(k)^2 \end{bmatrix} \tag{6.39}$$

where

$$\left|\mathbf{H}[\kappa]^T\mathbf{H}[\kappa]\right| = L\sum_{k=1}^L i(k)^2 - \left(\sum_{k=1}^L i(k)\right)^2 \tag{6.40}$$

From the above, the CRLB of estimating R_0 and V_o, the first and second diagonal elements, respectively, of (6.36) can be written as

$$\text{CRLB}(R_0) = \frac{\sigma_v^2}{\sum_{k=1}^L i(k)^2 - \frac{1}{L}\left(\sum_{k=1}^L i(k)\right)^2} \tag{6.41}$$

$$\text{CRLB}(V_o) = \frac{(\sigma_v^2/L)\sum_{k=1}^L i(k)^2}{\sum_{k=1}^L i(k)^2 - \frac{1}{L}\left(\sum_{k=1}^L i(k)\right)^2} \tag{6.42}$$

$$\text{CRLB}(V_o) = \left(\frac{\sigma_v^2}{L}\right)\left(\frac{1}{1 - \frac{1}{L}\left(\frac{\left(\sum_{k=1}^L i(k)\right)^2}{\sum_{k=1}^L i(k)^2}\right)}\right) \tag{6.43}$$

In other words, one can write

$$E\left((\hat{R}_0 - R_0)^2\right) \geq \text{CRLB}(R_0) \tag{6.44}$$

$$E\left((\hat{V}_o - V_o)^2\right) \geq \text{CRLB}(V_o) \tag{6.45}$$

Let us first consider $\text{CRLB}(R_0)$ in (6.41). The estimation accuracy depends on the following three factors:

- Measurement noise variance σ_v^2: The lower the measurement noise, the lower the CRLB.

- Number of observations L: Under the given assumptions, that R_0 remains a constant, more measurements will decrease estimation error.

- Current profile $i(k)$, $k = 1, \ldots, L$: The current profile should be selected in a way that the error bound in (6.41) is minimized.

Out of the three factors influencing the estimation error of R_0, two are constants. The current profile $i(k)$, $k = 1, \ldots, L$ should be selected in a way that the error can be made as small as possible.

Remark 6.1 Let us assume all the values of current are the same (i.e., $i(1) = i(2) = \ldots = i(L)$). This will make the denominator of (6.41) zero and lead to infinite error variance. Another way to look at it is that all equal values of $i(k)$ will make $\mathbf{A}$ rank deficient.

We need to find the current profile $i(1), i(2), \ldots, i(L)$ such that the $\text{CRLB}(R_0)$ can be reduced. The problem can be formally stated as follows:

For a given number of measurements L find $i(1), i(2), \ldots, i(L)$ such that the following cost function is maximized:

$$\mathcal{J}_{R_0}(i(1), i(2), \ldots, i(L)) = \sum_{k=1}^{L} i(k)^2 - \frac{1}{L}\left(\sum_{k=1}^{L} i(k)\right)^2 \tag{6.46}$$

under the constraint that

$$i_{\min} \leq i(1), i(2), \ldots, i(L) \leq i_{\max} \tag{6.47}$$

It can be shown that, for given values of the current limits $i_{\min}$ and $i_{\max}$, a current profile that alternates between the two extreme values will minimize $\mathrm{CRLB}(R_0)$. Minimization of $\mathrm{CRLB}(V_o)$, can be formulated as the following problem:

For a given number of measurements L, find $i(1), i(2), \ldots, i(L)$ such that the following cost function is minimized:

$$\mathcal{J}_{V_o}(i(1), i(2), \ldots, i(L)) = \left(\frac{\left(\sum_{k=1}^{L} i(k) \right)^2}{\sum_{k=1}^{L} i(k)^2} \right) \tag{6.48}$$

under the constraint that

$$i_{\min} \leq i(1), i(2), \ldots, i(L) \leq i_{\max} \tag{6.49}$$

Selecting current limits such that

$$i_{\min} = -i_{\max} \tag{6.50}$$

and a current profile that alternates between $i_{\min}$ and $i_{\max}$ will minimize both $\mathrm{CRLB}(R_0)$ and $\mathrm{CRLB}(V_o)$. Here, it is important to note that such a selection will result in

$$\sum_{k=1}^{L} i(k) = 0 \tag{6.51}$$

This condition can be satisfied in many different ways; two of which are listed below:

1. Adjacent $i(k)$ values that alternate between $i_{\min}$ and $i_{\max}$, that is, $i(1) = i_{\min}$, $i(2) = i_{\max}$, $i(3) = i_{\min}, \ldots$.

2. The first half of $i(k)$ is $i_{\min}$ and the second half is $i_{\max}$, that is, $i(1) = i(2) = \ldots i(L/2) = i_{\min}$ and $i(L/2 + 1) = i(L/2 + 2) = \ldots i(L) = i_{\max}$.

The current profiles discussed in items (a) and (b) are discrete; that is, the amplitude of the current is either $i_{\min}$ or $i_{\max}$. Generating such discrete current profiles may face additional hardware requirements in practical applications. Sinusoidal excitation signals can be relatively easily generated in practical systems. Let us consider a sample current profile $i(k)$ that is sampled from a sinusoid such that $i(k) = a \sin(2\pi k/n)$; here, each period of the sinusoid has n samples and a is the amplitude of the signal. Figure 6.2 shows an example of a sinusoidal signal for $n = 25, L = 100$, and $a = 1$. The sampled discrete values of $i(k)$ are shown by markers "o." This signal has $L/n = 100/25 = 4$ complete cycles. It is also important to note that the samples satisfy (6.51).

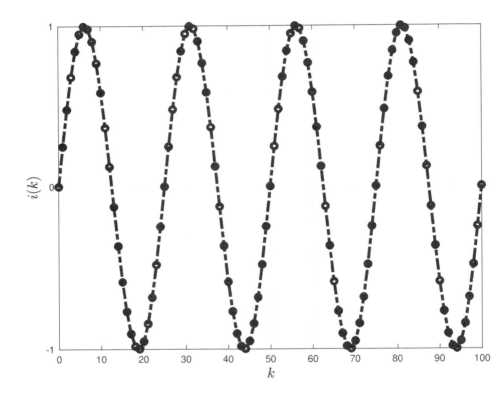

Figure 6.2 Sinusoidal current excitation signal.

Example 6.1

Consider a current excitation signal where adjacent $i(k)$ values that alternate between 1A and -1A (i.e., $i(1) = 1$, $i(2) = -1$, $i(3) = 1$, ..., $i(100) = -1$). Assume the voltage measurement error to be zero-mean with standard deviation $\sigma_v = 0.1$A.

- Compute the s.d. of the resistance estimation error.

- Assume that a sinusoidal current is described as $i(k) = a\sin(2\pi k/n)$ where $a = 1$ and $n = 25$. How many cycles of data are needed to achieve the same s.d. of the resistance estimation error?

- According to (6.41), the resistance estimates will have the following variance

$$\text{CRLB}_a(R_0) = \frac{0.1^2}{\sum_{i=1}^{100} 1^2} = 0.0001 \qquad (6.52)$$

The s.d. of the internal resistance estimation error is then $\sqrt{0.0001} = 0.01\Omega = 10\,\text{m}\Omega$.

- For one cycle of the sinusoidal signal, the resistance estimation error variance (6.41) will be

$$\text{CRLB}_{b1}(R_0) = \frac{0.1^2}{\sum_{k=1}^{25}\sin(2\pi k/25)} = 0.0008 \qquad (6.53)$$

That is, $\text{CRLB}_b(R_0) = 8\text{CRLB}_a(R_0)$; that means that eight cycles of sinusoidal data will be required to achieve the same level of accuracy achieved by $L = 100$ data points. That is,

$$\text{CRLB}_{b8}(R_0) = \frac{0.1^2}{\sum_{k=1}^{200}\sin(2\pi k/25)} = 0.0001 \qquad (6.54)$$

□

The performance analysis presented in this section shows that the accuracy of the estimation depends on the excitation signal. By carefully selecting the excitation signal, the accuracy of ECM parameter estimation can be improved. When there is no control over the excitation signal (e.g., when using battery usage data for ECM parameter estimation), the CRLB provides the lower bound on the ECM parameter estimation error.

Remark 6.2 The analysis in this section was done under the assumption that the current is the excitation signal and the voltage is the measured one; as such, it is assumed that the current $i(k)$ is perfectly known and that only the voltage suffers from measurement noise. However, there may be uncertainties in the current as well; these uncertainties are associated with the errors in generating the exact excitation waveform. When there is uncertainty in the current, the derivation and analysis of this section must be redone.

6.6 SIMULATION ANALYSIS

The data for the demonstration in this section was generated using a battery simulator introduced in Chapter 3. Figure 6.3 shows the battery simulator in the form of a block diagram. The battery simulator uses the equivalent circuit model shown in Figure 6.1 to simulate the voltage and current measurements that resemble real-time measurements from a battery. The OCV effect of the battery, denoted by $v_0[k]$ in Figure 6.1, was generated using the combined+3 model [2] with the following model parameters: $k_0 = -9.082$, $k_1 = 103.087$, $k_2 = -18.185$, $k_3 = 2.062$, $k_4 = -0.102$, $k_5 = -76.604$, $k_6 = 141.199$, and $k_7 = -1.117$. The voltage measurements across the battery were simulated using the observation model in (6.3). The voltage and current measurement noises were implemented based on (6.2) and (6.1), respectively, where the standard deviations of the voltage and current measurement noise were assumed to be equal in magnitude; that is, $\sigma_v = \sigma_i = \sigma$, where σ was computed based on the assumed signal-to-noise ratio of the measurement system, defined as

$$\text{SNR} = 20 \log\left(\frac{I}{\sigma}\right) \tag{6.55}$$

where $I = |i(k)|$, $k = 1, \ldots, L$ is the amplitude of the current signal that is assumed to be constant throughout the entire simulation. The relaxation parameters of the ECM are set at $R_0 = 0.2$, $R_1 = 0.3$, $C_1 = 50$, $R_2 = 0.3$, and $C_2 = 500$. The electrical ECM in the battery simulator can be changed in a way that the RC models can be selected, from the set of $\{(R_0), (R_1, C_1), (R_2, C_2)\}$.

First, let us consider the current profile shown in Figure 6.4. The current is sampled at $10\,\text{Hz}$, resulting in $L = 1000$ samples. The current profile $i(1), i(2), \ldots, i(L)$ also holds the property (6.51) where $L = 1000$.

Figure 6.5 shows a plot of OCV $v_o(k)$ over time. It can be seen that when the current $i(k)$ is positive, $v_o(k)$ increases, and when $i(k)$ is negative, $v_o(k)$ decreases. Since the average current shown in Figure 6.4 is zero, the average OCV in Figure 6.5 is constant as well. It is also important to note that the OCV is not affected by

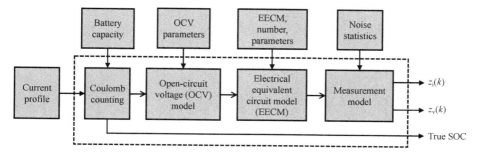

Figure 6.3 Battery simulator. Reproduced with permission from [3].

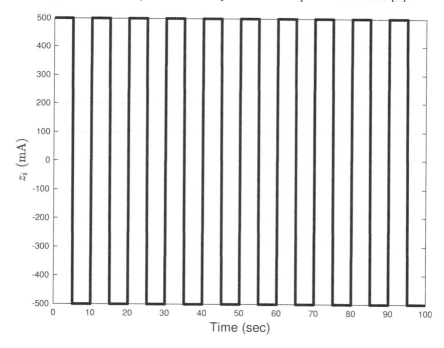

Figure 6.4 Current profile 1. Reproduced with permission from [3].

hysteresis/relaxation effects of the battery; regardless of the ECM assumption, the OCV in Figure 6.5 will be the same for the given current profile. However, the ECM affects the terminal voltage $v(k)$.

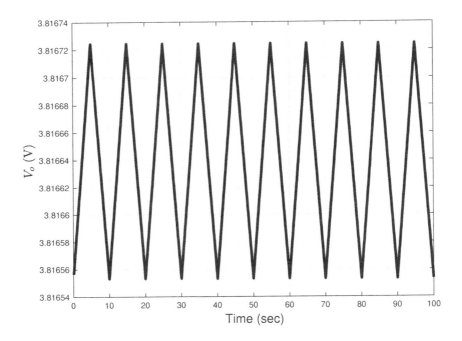

Figure 6.5 Simulated OCV. The average OCV is $V_o = 3.8165649$V. Reproduced with permission from [3].

Figure 6.6 shows a plot of the true voltage across the battery terminals, $v(k) = v_o(k) + i(k)R_0$, over time. It must be noted that even though $v_o(k)$ changes with time, the magnitude of the voltage drop $i(k)R_0$ remains a constant. Also, that the magnitude of change in $v_o(k)$ (see Figure 6.5) is relatively insignificant compared to the magnitude of $i(k)R_0$. Consequently, the magnitude of $v(k)$ appears unchanged in Figure 6.6. Another explanation for this observation is that the change in OCV is small, within a duration of 5 seconds.

Figure 6.7 shows the (simulated) voltage measurements from the battery simulator. Here, the battery simulator is set to Model 2 ECM (see Figure 6.1) (i.e., only $R_0 = 0.2\Omega$ had a nonzero value and all other ECM parameters were set to zero). For now, it is

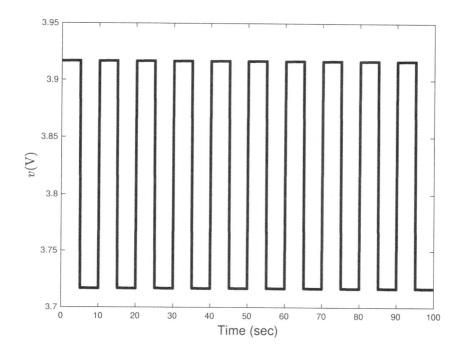

Figure 6.6 Simulated voltage measurements. Reproduced with permission from [3].

assumed that the current profile is perfectly known, as shown in Figure 6.4 (i.e., it is assumed that the current measurement noise is zero).

The least-square estimation algorithm (6.34) for Model 2 was used to estimate the resistance R_0 and V_o. Let us denote these estimated quantities as $\hat{R}_0$ and $\hat{V}_o$, respectively. The normalized mean square error (NMSE) of these estimates is defined as

$$\text{NMSE}(R_0) = \frac{1}{R_0^2} \sum_{m=1}^{M} (R_0 - \hat{R}_0(m))^2 \tag{6.56}$$

$$\text{NMSE}(V_o) = \frac{1}{V_o^2} \sum_{m=1}^{M} (V_o - \hat{V}_o(m))^2 \tag{6.57}$$

where M denotes the number of Monte Carlo runs.

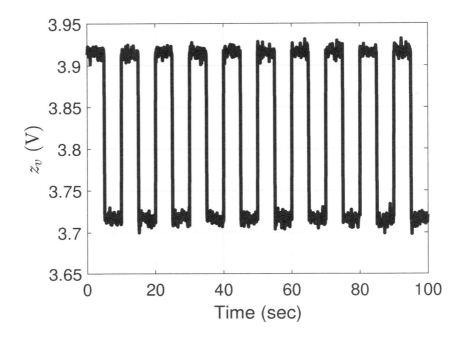

Figure 6.7 Simulated voltage measurements (Model 2). Reproduced with permission from [3].

In order to make the CRLB comparable to the NMSE defined in (6.56) and (6.57), the following CRLB values in (6.41) and (6.43) were computed for comparison during simulation studies.

$$\text{CRLB}(R_0) \rightarrow \frac{\text{CRLB}(R_0)}{R_0^2} \tag{6.58}$$

$$\text{CRLB}(V_o) \rightarrow \frac{\text{CRLB}(V_o)}{V_o^2} \tag{6.59}$$

6.6.1   Perfect ECM Assumption

In a perfect ECM assumption, the battery management system assumes the same model as the battery simulator. We will now consider a scenario where the battery simulator and BMS assume Model 2. The NMSE for R_0 is calculated using (6.56), where $\hat{R}_0$ for

Model 2 is estimated over 1,000 Monte Carlo runs. The CRLB for R_0 is calculated using (6.58), where the length of current samples (Figure 6.4) is $L = 1,000$. Figure 6.8 shows the NMSE and CRLB for R_0 estimation under the model-matched assumption. It can be observed that the NMSE is close to the theoretical bound CRLB for all SNR levels indicating an efficient estimator.

Similarly, in Figure 6.9, the NMSE of V_o estimate is plotted with CRLB of V_o for SNR values between 0 and 40 dB. The NMSE of V_o estimate is calculated using (6.57) and the CRLB of V_o estimate is calculated using (6.59). It can be observed again that the performance of the estimator is close to the theoretical bound CRLB and that the estimator is efficient.

Now let us compare the resistance estimates under the perfect ECM assumption for a particular SNR level. Table 6.1 contains the R_0 estimate for Model 2 under perfect ECM assumption at SNR = 20 dB. When the BMS and battery simulator assumes Model 2, the estimate of resistance R_0 is obtained as an average from the estimate of R_0 over 1,000 Monte Carlo runs. The table also shows another case of perfect ECM assumption where the simulator and the estimator correspond to Model 3. Due to this assumption, the model parameter vector consists of two resistance R_0 and R_1 values as in Figure 6.1(c). Thus, when the BMS and battery simulator assume Model 3, the estimate of resistances, R_0 and R_1 are obtained as an average from the estimates of R_0 and R_1 over 1,000 Monte Carlo runs.

Table 6.1
Average Estimates of the ECM Parameters (20 dB)

Battery Simulator	BMS	R_0	R_1	$\hat{R}_0$	$\hat{R}_1$
Model 2	Model 2	0.2	NA	0.1999	NA
Model 3	Model 3	0.2	0.1	0.2114	0.0740

6.6.2 Realistic ECM Assumption

In this section, a simulation-based ECM parameter estimation analysis is presented based on a realistic ECM assumption in which the battery management system assumes an ECM model that is different from the one used by the battery simulator to simulate the measurements. In this section, we consider a scenario where the battery simulator assumes Model 3 and the BMS assumes Model 2. It must be noted that the CRLB derivations are done under perfect model assumptions. Figure 6.10 shows the NMSE and CRLB computed under the model mismatch assumption. The NMSE is significantly

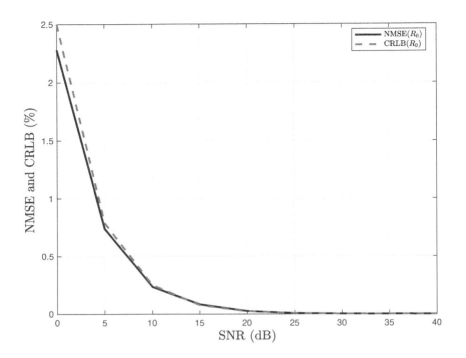

Figure 6.8 Performance of R_0 estimation using ECM Model 2. Reproduced with permission from [3].

greater than CRLB at all SNR levels. An explanation for this observation can be stated based on the model assumptions made for this simulation. The ECM Model 3 contains two resistor components, R_0 and R_1. When a lower-order model, here Model 2, is used to estimate the parameters the resulting estimate of the resistance is observed to be closer to the sum of the two resistor components of Model 3. To confirm this observation, let us define a new type of NMSE as follows:

$$\text{NMSE}(R_{\text{tot}}) = \frac{1}{R_{\text{tot}}^2} \sum_{m=1}^{M} (R_{\text{tot}} - \hat{R}_0(m))^2 \qquad (6.60)$$

Here the estimation error is computed with respect to the total resistance, defined as $R_{\text{tot}} = R_0 + R_1$. Figure 6.11 shows the computed NMSE based on the two different definitions given in (6.56) and (6.60). In this figure, it can be observed that $\text{NMSE}(R_{\text{tot}})$

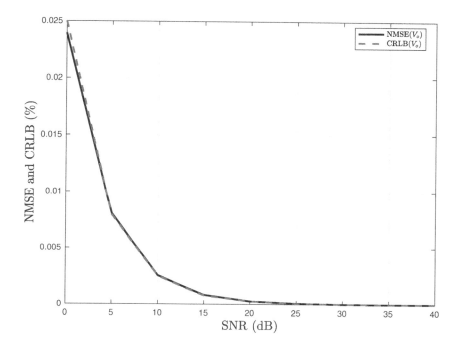

Figure 6.9 Performance of OCV estimation using ECM Model 2. Reproduced with permission from [3].

is less than the NMSE for $\text{NMSE}(R_0)$ at all SNR levels. This means that the BMS under the Model 2 assumption estimates both resistances together, as a summation. Thus, it can be confirmed that the estimation of resistance of an ECM Model 2 is approximately the sum of the two resistor components of Model 3.

Now let us compare the estimates of resistances under the perfect and realistic ECM assumptions at a particular SNR level. Table 6.2 contains the averages of the R_0 and R_1 estimates for both assumptions of ECM Model 3. The parameter estimate is obtained as an average from the estimates of 1,000 Monte Carlo runs at SNR = 20 dB. While using the realistic ECM, the estimation algorithm assumes a different model, Model 2. Thus, from the ECM in Figure 6.1(b), one estimate of resistance is obtained (i.e., 0.2823Ω). Now we apply the observation that the resistance of an ECM Model 2 is approximately the sum of the two resistor components of Model 3. Conforming to this observation, the Model 2 estimate is shown to be the total resistance of ECM Model 3 $0.2114 + 0.074 \approx 0.2823$ under perfect ECM assumption.

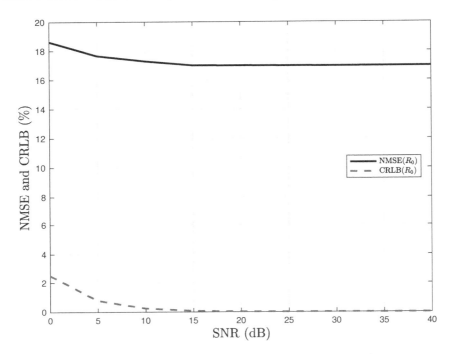

Figure 6.10 Performance of R_0 estimation. Reproduced with permission from [3].

Table 6.2
Average Estimates of ECM Parameters (20 dB)

Battery Simulator	BMS	R_0	R_1	$\hat{R}_0$	$\hat{R}_1$
Model 3	Model 3	0.2	0.1	0.2114	0.0740
Model 3	Model 2	0.2	0.1	0.2823	NA

The simulation analysis presented in this section under realistic ECM assumption is important because of the fact that when it comes to real battery applications, the

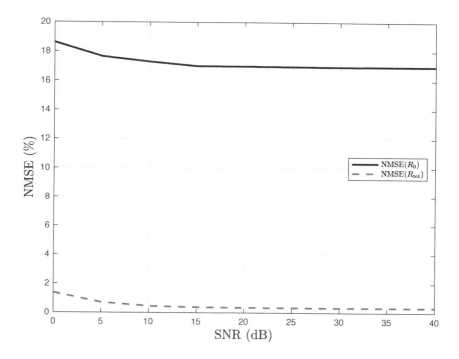

Figure 6.11 Performance of R_0 estimation. Reproduced with permission from [3].

assumed model is always different from the realistic case. Also, most battery management systems resort to reduced-order models in order to save computation and hardware complexity.

6.6.3 Real Data

In this section, the performance of the proposed approach for battery parameter estimation using data collected from a Samsung-30T INR21700 battery cell is presented. A current profile, shown in Figure 6.12 is applied to the battery and the voltage across the battery terminals is recorded. The data collection is performed using an Arbin BT-2000 battery cycler. The data presented here is openly available in Mendeley Data at 10.17632/h3yfxtwkjz.1, https://data.mendeley.com/datasets/h3yfxtwkjz. The s.d. of the voltage and current measurement error of the device is approximately 0.00033V and 0.00025A, respectively. The sampling time of the data is $\Delta = 1$ second; this resulted in close to 7,200 voltage and current measurements, as shown in Figure 6.12. Then, least-square estimation (6.34) is performed on the recorded data to estimate the model parameters. Table 6.3 shows the parameters obtained when the estimation algorithm is set to ECM Model 2 and Model 3, respectively. The parameter estimate is obtained as an average from the estimates over 1,000 Monte Carlo runs. Here, while using Model 2 to estimate the parameters, the resistance estimate is $\hat{R}_0 = 0.0152\Omega$. When Model 3 is used, the individual resistance estimates are $\hat{R}_0 = 0.0107$ and $\hat{R}_1 = 0.0049$, resulting in $\hat{R}_0 + \hat{R}_1 = 0.0152$, which is approximately equal to the Model 2 estimation of $\hat{R}_0$. Thus, the observation that while using Model 2, the resistance obtained is closer to the summation of all the resistor components of Model 3 holds true for real battery data.

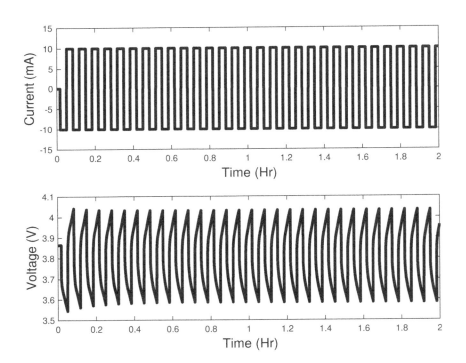

Figure 6.12 Real current and voltage measurements. Reproduced with permission from [3].

Table 6.3

Estimates of ECM Parameters (Real Data)

BMS	$\hat{R}_0$	$\hat{R}_1$
Model 2	0.0152	NA
Model 3	0.0107	0.0049

6.7 SUMMARY

The electrical ECM of a battery represents its behavior in response to voltage and current excitations. Typical ECMs consist of resistors, capacitors, inductors, and hysteresis voltage models. Unlike the OCV-SOC parameters, details shown in Chapter 3, the parameters of the ECM change in response to temperature changes, SOC, and calendar age. Hence, the ECM parameters need to be estimated in real time. Estimated ECM parameters are needed for SOC estimation, state of power estimation, remaining mileage estimation, and SOH estimation.

ECM parameter estimation requires special excitation signals to be applied to the battery. For instance, an RC element with a very large time constant will require very low-frequency excitation signals. That is, it takes longer to estimate RC elements with a long time constant. Similarly, the inductive responses from a battery are observable only at very high frequencies. Most ECM parameters relate nonlinearly to voltage/current observations. This chapter outlined an approach that approximates these nonlinearities into a linear observation model. Further, in real-time scenarios, the nature of the current profile can be exploited to estimate certain ECM parameters more accurately than others.

It is shown in this chapter that the achievable accuracy in ECM parameter estimation depends on the excitation signals used. The achievable accuracy is quantified in terms of the CRLB, which is the lower bound on the estimation error variance. It is shown that the CRLB analysis can be used to develop optimized excitation signals for faster and more accurate ECM parameter estimation.

6.8 BIBLIOGRAPHICAL NOTES

There is a vast literature about time-domain approaches to battery ECM identification. However, most of the existing ECM identification techniques are tied to SOC estimation. A very early and relevant example of this approach can be found in [4]. Frequency-domain approaches to independent ECM identification have been widely reported. Chapter 5 was dedicated to frequency-domain approaches to ECM identification. A more detailed discussion about the effect of noise in both current and voltage measurements is presented in [5].

References

[1] Y. Bar-Shalom, X.R. Li, and T. Kirubarajan, *Estimation with Applications to Tracking and Navigation: Theory, Algorithms and Software*, John Wiley & Sons, New York, 2004.

[2] B. Pattipati, B. Balasingam, G. Avvari, K. R. Pattipati, and Y. Bar-Shalom, "Open circuit voltage characterization of lithium-ion batteries," *Journal of Power Sources*, Vol. 269, pp. 317-333, 2014.

[3] P. Pillai, S. Sundaresan, K.R. Pattipati, and B. Balasingam, "Optimizing current profiles for efficient online estimation of battery equivalent circuit model parameters based on Cramer-Rao lower bound," *Energies*, Vol. 15, No. 22, pp. 8441, 2022.

[4] G. L. Plett, *Battery Management Systems, Volume II: Equivalent-Circuit Methods*, Artech House, Norwood, MA, 2015.

[5] B. Balasingam, and K.R. Pattipati, "On the identification of electrical equivalent circuit models based on noisy measurements," *IEEE Transactions on Instrumentation and Measurement*, Vol. 70, pp. 1-16, 2021.

Chapter 7

Battery Capacity Estimation

7.1 INTRODUCTION

From the moment a battery is made, it starts to go through a process known as the capacity fade. Capacity fade is caused by aging, fast charging, heavy usage, and extreme environmental conditions. Robust battery management needs accurate knowledge of the battery capacity to estimate all critical states required for effective battery management: state of charge, state of power, state of health, time to empty, and remaining useful life. Incorrect knowledge of battery capacity may lead to consequential decisions, such as overcharging, that may jeopardize the safety of the energy storage system.

The simplest approach to tracking the battery capacity is to use a capacity fade model. Figure 7.1 shows the battery capacity against the number of cycles obtained over 500 identical charge-discharge cycles. This data can be fitted to a model to obtain a capacity fade model.

The problem with the capacity fade model in Figure 7.1 is that the data is from identical charge-discharge cycles that were collected consecutively without any breaks; the charge/discharge currents were the same in all cycles, the battery was kept at the temperature all the time, and there were no nonuniform time gaps between cycles. Real-world conditions could be less uniform: there will be uneven gaps between cycles, the length of charge/discharge cycles will not be complete, and the environmental conditions will not be fixed.

This chapter explains approaches to estimating the battery capacity in real-time without the use of a capacity fade model. The approach presented in this chapter relies on the relative stability (over time) of OCV-SOC curves for capacity estimation in real-time.

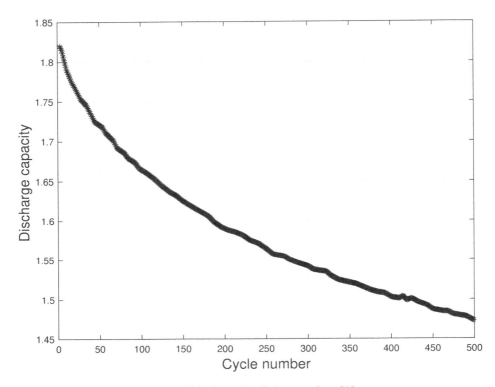

Figure 7.1 Capacity fade over time [1].

7.2 BASICS OF BATTERY CAPACITY ESTIMATION

7.2.1 Offline Estimation of Battery Capacity

In an offline setting, the battery capacity can be estimated by completely discharging the battery from when it was full until it becomes empty. By denoting the OCV of the battery as OCV_{max} and OCV_{min} when it is full and empty, respectively, it can be said that the total capacity of the battery can be estimated by discharging the battery from $OCV = OCV_{max}$ to $OCV = OCV_{min}$. However, such complete discharge is only possible with an infinitely small current. That is, computing the total capacity is a time-consuming process. In typical OCV characterization experiments, an approximate value of the total capacity is computed by discharging the battery at the C/30-rate, taking approximately 30 hours to compute the total capacity.

Rated capacity measurements can be obtained relatively fast. The C-rate capacity is estimated by discharging the battery at 1C-rate from full, that is, when the rested voltage measured $OCV = OCV_{max}$ until the terminal voltage reaches $OCV = OCV_{min}$. The experimental time for the 1C rate capacity estimation will last slightly less than 1 hour.

The C-rate is a commonly accepted way of indicating the charge/discharge current for a battery. For example, let us consider a battery with a rated capacity of 3 Ah (or 3000 mAh). When this battery is said to be discharged at a C/30 rate, the discharge current is $3000/30 = 100$ mA. Similarly, when this battery is said to be charged at a 1C rate, the charge current is 3A. It takes approximately 30 hours to discharge a battery from full to empty at C/30-rate discharge. However, when charging a battery at a 1C rate, the voltage drop becomes significant and prevents the battery from fully charging; for more details on this, please read Chapter 10.

The battery capacity fades over time depending on age, environmental conditions, and usage patterns. It is important to keep track of the battery capacity to have accurate knowledge of the state of health of the battery. That is, there is a critical need to estimate the battery capacity in real-time.

7.2.2 Real-Time Capacity Estimation

The open-circuit voltage model of a battery can be exploited to estimate the battery capacity in real time. Consider the scenario illustrated in Figure 7.2 where the battery measured $OCV = OCV_1$ at the start of the experiment. The battery is then discharged by extracting C Coulombs (measured in Ah) from it. At the end of this discharge, and after sufficiently resting the battery, the battery measured $OCV = OCV_2$.

Let us assume that the following combined+3 model is used to represent the OCV-SOC characterization of the battery

$$OCV = f_{OCV}(s) = k_0 + \frac{k_1}{s} + \frac{k_2}{s^2} + \frac{k_3}{s^3} + \frac{k_4}{s^4}$$
$$+ k_5 s + k_6 \ln(s) + k_7 \ln(1 - s) \tag{7.1}$$

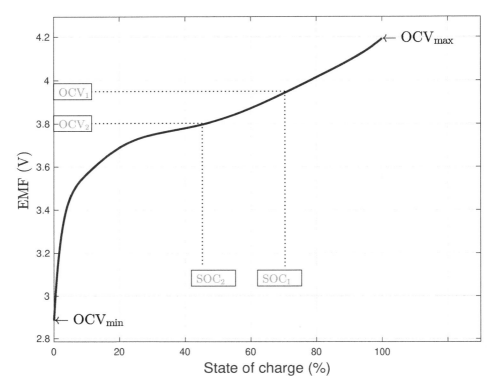

Figure 7.2 OCV-based capacity estimation.

where s denotes the SOC. Assuming that the parameters $k_0, \ldots, k_7$ of the OCV-SOC model are known, the SOC can be computed for a given OCV, that is,

$$s = f_{\text{OCV}}^{-1}(\text{OCV}) \tag{7.2}$$

Now, for the two OCV measurements in Figure 7.2, the corresponding SOC values can be obtained as

$$\text{SOC}_1 = f_{\text{OCV}}^{-1}(\text{OCV}_1) \tag{7.3}$$

$$\text{SOC}_2 = f_{\text{OCV}}^{-1}(\text{OCV}_2) \tag{7.4}$$

The change in SOC is equal to the change in Coulombs normalized by the battery capacity, that is,

$$d_{SOC} = SOC_2 - SOC_1 = \frac{C}{Q} \tag{7.5}$$

where C denotes the change in Coulombs and Q denotes the battery capacity. Using the relationship (7.5), the battery capacity can be estimated. This is the fundamental concept for battery capacity estimation [2, 3]. Several practical BMS products, such as the commercial battery fuel gauge [4] by Texas Instruments.

In a real-world setting, the OCV measurements can be corrupted by various sources of noise. The remainder of this chapter details various filtering approaches for the accurate real-time estimation of battery capacity in the presence of noise.

7.3 CAPACITY ESTIMATION IN THE PRESENCE OF NOISE

A typical scenario involving noisy OCV measurement is shown in Figure 7.3. The solid line shows the true OCV curve and the dashed line shows the envelope of error. This error could be due to hysteresis and relaxation effects within the battery. To illustrate this further, consider the ECM of the battery, shown in Figure 7.4. The relaxation effect of the battery is modeled as a series of RC circuits; when the battery is fully at rest, the relaxation effect becomes zero (i.e., when there is no current activity through the battery for sufficient time, the voltage across the RC circuits becomes zero) [5]: "The hysteresis in Li-ion batteries is generated due to the thermodynamic entropic effects, mechanical stress, and microscopic distortions within the active electrode materials during lithium insertion/extraction." The hysteresis effect does not vanish when resting the battery. When the battery is fully rested, the relaxation error becomes zero. The measured voltage at time t can be written as

$$z_v[t] = V_o(s[t]) + h[t] \tag{7.6}$$

where the hysteresis $h[t]$ corrupts the measured OCV.

It is likely that the hysteresis effect is significantly higher after charging than it is after discharging. Such details could be incorporated in advanced hysteresis models, such as the enhanced self-correcting model reviewed in Chapter 3. In this chapter, it is assumed that the effect of hysteresis is similar following both charging and discharging currents.

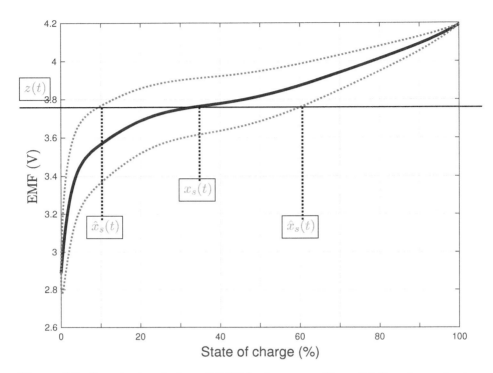

Figure 7.3 Generic description of OCV lookup error. The solid line shows the true OCV-SOC characteristics. The dashed lines describe the effect of hysteresis when the battery comes to rest after charging (above the solid line) or discharging (below the solid line).

The OCV-SOC characteristic curve can be used to get a measure of SOC whenever the battery is sufficiently rested. The SOC of the battery for a given at-rest terminal voltage (which is also the open-circuit voltage) $z_v[t]$, written as

$$\hat{x}_s[t] = f_{\text{OCV}}^{-1}(z_v[t]) \tag{7.7}$$

can be computed using the OCV-SOC characterization by computing the inverse of (7.1). There are several methods for computing the inverse of a nonlinear function, such as Newton's method and binary search [6].

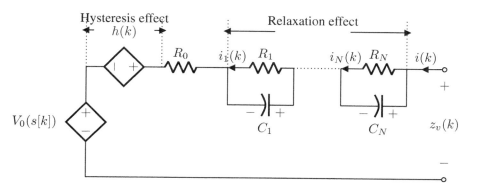

Figure 7.4 Hysteresis and relaxation effects in a battery. The relaxation effect can be determined after estimating the equivalent circuit model parameters of the battery. Accurate hysteresis estimation remains a challenging problem.

The SOC estimation error in (7.7) is modeled as follows

$$x_s[t] = \hat{x}_s[t] + \tilde{x}_s[t] \tag{7.8}$$

where the OCV lookup error $\tilde{x}_s[t]$ is caused by the hysteresis effect as illustrated in Figure 7.3. Figure 7.3 shows that when the battery comes to rest at time t after a discharging process, the OCV lookup error $\tilde{x}_s[t]$ will be negative. Similarly, when the battery comes to rest at time t after a charging process, the OCV lookup error $\tilde{x}_s[t]$ will always be positive. However, the magnitude of the error will vary with the amount of hysteresis, which is a function of the magnitude of the current before rest, SOC, and time.

Let us assume that the SOC_1 and SOC_2 in (7.5) are measured at rest time t_1 and t_2, respectively. Let us denote these as

$$x_s[t_1] \triangleq SOC_1 \tag{7.9}$$

$$x_s[t_2] \triangleq SOC_2 \tag{7.10}$$

According to the model in (7.8), the above SOC estimates can be written as

$$x_s[t_1] = \hat{x}_s[t_1] + \tilde{x}_s[t_1] \tag{7.11}$$

$$x_s[t_2] = \hat{x}_s[t_2] + \tilde{x}_s[t_2] \tag{7.12}$$

where $\hat{x}_s[t_1]$ and $\hat{x}_s[t_2]$ refer to the computed SOC according to (7.7) and $\tilde{x}_s[t_1]$, $\tilde{x}_s[t_2]$ refer to the SOC estimation error due to hysteresis.

By subtracting (7.11) from (7.12) and making use of the relationship in (7.5), we can write

$$d_s[k] = Q^{-1}d_c[k] + \tilde{w}_s[k] \tag{7.13}$$

where

$$d_c[k] = C \tag{7.14}$$
$$d_s[k] = \hat{x}_s[t_2] - \hat{x}_s[t_1] \tag{7.15}$$
$$\tilde{w}_s[k] = \tilde{x}_s[t_1] - \tilde{x}_s[t_2] \tag{7.16}$$

It must be noted that, regardless of the fact that the sign of OCV lookup error $\tilde{x}_s[t]$ is biased towards the battery mode $\in \{$charging, discharging$\}$, the differential error $\tilde{w}_s[k]$ (defined in (7.16)) can either be positive or negative. By considering a large number of differential errors, we assume that $\tilde{w}_s[k]$ is approximately white.

In this chapter, we assume that the OCV-SOC curve is obtained through the normalized OCV modeling approach presented in Chapter 4. For such OCV-SOC modeling, the data is collected at a very slow rate of charging and discharging; the resulting OCV-SOC curve is free of charging or discharging effects. With such an OCV-SOC curve, the assumption that $\tilde{w}_s^k$ is white becomes closer to reality.

Consider that there are N rest states in a batch of measurements; this will result in $N(N-1)/2$ measurements similar to (7.13). For example, for $N = 4$, let us assume that the battery is in rest state at time instances t_1, t_2, t_3 and t_4. Table 7.1 shows the possible rest pairs for differential observations.

Let us assume that a κth batch of L measurements of (7.13) are made. This can be represented as

$$\mathbf{d}_s[\kappa] = Q^{-1}\mathbf{d}_c[\kappa] + \tilde{\mathbf{w}}_s[\kappa] \tag{7.17}$$

where

$$\mathbf{d}_s[\kappa] = \left[d_s^1, d_s^2, \ldots, d_s^L\right]^T \tag{7.18}$$
$$\mathbf{d}_c[\kappa] = \left[d_c^1, d_c^2, \ldots, d_c^L\right]^T \tag{7.19}$$
$$\tilde{\mathbf{w}}_s[\kappa] = \left[w_s^1, w_s^2, \ldots, w_s^L\right]^T \tag{7.20}$$

In the practical usage of mobile devices, there will be many time epochs of rest states within each cycle of usage. It must be noted that the batch length L can be time-varying.

Table 7.1

Possible Rest Pairs for OCV-Based Capacity Estimation

First Rest	Second Rest	Measurement Index (k)
t_1	t_2	$k = 1$
t_1	t_3	$k = 2$
t_1	t_4	$k = 3$
t_2	t_3	$k = 4$
t_2	t_4	$k = 5$
t_3	t_4	$k = L = 6$

7.3.1 LS Estimate

Now the LS estimate of the inverse battery capacity is given by

$$\hat{Q}_{\mathrm{LS}}^{-1} = \left((\mathbf{d}_c[\kappa])^T (\boldsymbol{\Sigma}_{\tilde{\mathbf{w}}_s}[\kappa])^{-1} \mathbf{d}_c[\kappa] \right)^{-1} (\mathbf{d}_c[\kappa])^T (\boldsymbol{\Sigma}_{\tilde{\mathbf{w}}_s}[\kappa])^{-1} \mathbf{d}_s[\kappa] \tag{7.21}$$

where $\boldsymbol{\Sigma}_{\tilde{\mathbf{w}}_s}[\kappa]$ is the covariance matrix with noise $\tilde{\mathbf{w}}_s[\kappa]$.

$$\boldsymbol{\Sigma}_{\tilde{\mathbf{w}}_s}[\kappa] = E(\tilde{\mathbf{w}}_s[\kappa](\tilde{\mathbf{w}}_s[\kappa])^T) \tag{7.22}$$

For independent and identically distributed noise,

$$\boldsymbol{\Sigma}_{\tilde{\mathbf{w}}_s}[\kappa] = \sigma^2 \mathbf{I} \tag{7.23}$$

where $\sigma^2 = E[(\tilde{\mathrm{w}}_s^k)^2]$ is the variance of the noise, $\tilde{\mathrm{w}}_s^k$. The variance of the LS inverse capacity estimate is

$$R_{\mathrm{LS}}[\kappa] = \left((\mathbf{d}_c[\kappa])^T (\boldsymbol{\Sigma}_{\tilde{\mathbf{w}}_s}[\kappa])^{-1} \mathbf{d}_c[\kappa] \right)^{-1} \tag{7.24}$$

7.3.2 TLS Estimate

It must be noted that $\mathbf{d}_c^\kappa$ in (7.17) is constructed from measured current values that are known to be noisy, whereas the LS and RLS estimation methods assume that $\mathbf{d}_c^\kappa$ is perfectly known. For a more realistic solution, one should consider the uncertainty in $\mathbf{d}_c^\kappa$. In this section, we propose an approach based on total least-squares (TLS) optimization that addresses errors in both $\mathbf{d}_s^\kappa$ and $\mathbf{d}_c^\kappa$. For more details about the TLS method, the reader is referred to [7] and the references therein.

Let us construct the following augmented observation matrix

$$\mathbf{H}[\kappa] = \begin{bmatrix} \mathbf{d}_s[\kappa] & \mathbf{d}_c[\kappa] \end{bmatrix} \tag{7.25}$$

The information matrix associated with the augmented observation matrix is

$$\mathbf{S_H}[\kappa] = (\mathbf{H}[\kappa])^T \mathbf{H}[\kappa] \tag{7.26}$$

Let us write the eigendecomposition of $\mathbf{S_H^\kappa}$ in the following form

$$\mathbf{S_H}[\kappa] = \mathbf{V}[\kappa]\mathbf{\Lambda}[\kappa]\mathbf{V}[\kappa]^T \tag{7.27}$$

where $\mathbf{\Lambda}[\kappa]$ is a diagonal 2×2 matrix of non-negative eigenvalues arranged from the largest to the smallest, that is, $\mathbf{\Lambda}^\kappa(1,1)$ denotes the largest eigenvalue and $\mathbf{\Lambda}^\kappa(2,2)$ denotes the smallest eigenvalue. Each column of the 2×2 matrix $\mathbf{V}[k] = \begin{bmatrix} \mathbf{v}_1^\kappa, & \mathbf{v}_2^\kappa \end{bmatrix}$ has the corresponding eigenvectors, that is, the first column $\mathbf{v}_1^\kappa$ is the eigenvector corresponding to the largest eigenvalue and the second column $\mathbf{v}_2^\kappa$ is the eigenvector corresponding to the smallest eigenvalue.

The TLS estimate of the inverse battery capacity is then given by the ratio of the components of $\mathbf{v}_2^\kappa$, namely,

$$\begin{aligned} \hat{Q}_{\text{TLS}}^{-1}[\kappa] &= -\frac{\mathbf{v}_2^\kappa(1)}{\mathbf{v}_2^\kappa(2)} \\ &= \frac{\mathbf{S_H^\kappa}(1,2)}{\mathbf{S_H^\kappa}(1,1) - \Lambda^\kappa(2,2)} \end{aligned} \tag{7.28}$$

where $\mathbf{v}_2^\kappa \triangleq \mathbf{v}_2[\kappa]$ and (i,j) is used to denote the (i,j)th element of the vector/matrix and $\mathbf{S_H^\kappa}(i,j)$ is the (i,j)th element of $\mathbf{S_H}[\kappa]$. For more on the TLS approach to capacity estimation, the reader is referred to [1].

Remark 7.1 The TLS estimates are useful when the noise in the model is very high. It was shown in [8] that when the noise in the model is low, both the LS and TLS estimates coincide.

7.4   RECURSIVE ESTIMATES

The capacity estimates shown in (7.21) and (7.28) was based on a batch of at least two measurements. For example, the batch κ may refer to measurements taken on a certain day and the batch $\kappa + 1$ may refer to measurements taken from another day. The number of measurements in each batch may differ from one batch to another. The recursive approaches presented in this section can be used to update the capacity estimates based on batch measurements over a long time.

7.4.1 Recursive LS

When a new batch of $\{\mathbf{d}_s[\kappa + 1], \mathbf{d}_c[\kappa + 1]\}$ pair arrives, the LS estimates can be recursively updated by

$$P_{\mathrm{RLS}}^{-1}[\kappa + 1] = \lambda P_{\mathrm{RLS}}^{-1}[\kappa] + (\mathbf{d}_c[\kappa + 1])^T (\mathbf{\Sigma}_{\tilde{\mathbf{w}}_s}[\kappa + 1])^{-1} \mathbf{d}_c[\kappa + 1] \qquad (7.29)$$

$$\hat{Q}_{\mathrm{RLS}}^{-1}[\kappa + 1] = P_{\mathrm{RLS}}[\kappa + 1]\left(\lambda P_{\mathrm{RLS}}^{-1}[\kappa]\hat{Q}_{\mathrm{RLS}}^{-1}[\kappa] + (\mathbf{d}_c[\kappa + 1])^T (\mathbf{\Sigma}_{\tilde{\mathbf{w}}_s}^{\kappa+1})^{-1} \mathbf{d}_s[\kappa + 1]\right) \qquad (7.30)$$

where $P_{\mathrm{RLS}}^{-1}[\kappa]$ is the $L \times L$ information matrix for capacity estimation and λ is the fading memory constant. Figure 7.5 shows a block diagram of the RLS approach for recursive estimation of capacity.

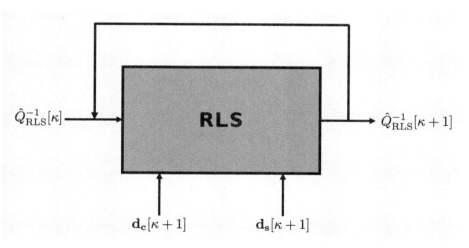

Figure 7.5 Block diagram of the RLS approach.

7.4.2   Recursive TLS

The information matrix in (7.26) can be updated with fading memory, as follows

$$\mathbf{S_H}[\kappa] = \lambda \mathbf{S_H}[\kappa - 1] + \frac{(\mathbf{H}[\kappa])^T \mathbf{H}[\kappa]}{L - 1} \tag{7.31}$$

Now, based on [9], the TLS estimation error covariance is (approximately)

$$R_{\text{TLS}}[\kappa] = \left(\frac{1}{(\mathbf{z}[\kappa])^T \mathbf{S_H}[\kappa] \mathbf{z}[\kappa]} \sum_{i=1}^{M} \mathbf{h}_i[\kappa] (\mathbf{h}_i[\kappa])^T \right)^{-1} \tag{7.32}$$

where $\mathbf{h}_i^\kappa$ is the ith row of $\mathbf{H}^\kappa$, M is the number of rows in $\mathbf{H}^\kappa$ and

$$\mathbf{z}[\kappa] = \left[-\frac{\mathbf{v}_2^\kappa(1)}{\mathbf{v}_2^\kappa(2)} \quad -1 \right]^T \tag{7.33}$$

Figure 7.6 shows a block diagram of the recursive TLS approach for recursive estimation of capacity.

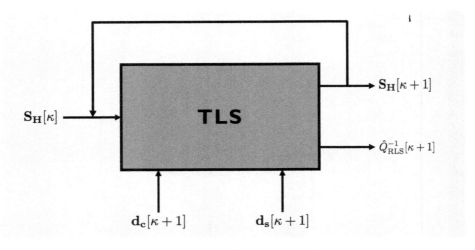

Figure 7.6 Block diagram of the recursive TLS approach.

7.4.3 KF-Based Fusion

The recursive approaches presented in Sections 7.4.1 and 7.4.2 both treat the capacity as a parameter that does not change over time. This section presents an approach to model the slow change in capacity over time and to track its change over time based on batch estimates.

Let us assume the battery capacity to be a random variable that undergoes the following slowly changing Wiener process

$$x_c[\kappa + 1] = x_c[\kappa] + w_c[\kappa] \tag{7.34}$$

where $w_c[\kappa]$ is assumed to be zero mean white Gaussian noise with variance $Q_c[\kappa]$. The capacity estimates, either $\hat{Q}_{\text{TLS}}[\kappa]$ or $\hat{Q}_{\text{LS}}[\kappa]$, fit the following observation model

$$z_c[\kappa] = x_c[\kappa] + n_c[\kappa] \tag{7.35}$$

where $n_c[\kappa]$ is assumed to be a zero mean white noise with variance

$$R_c[\kappa] = \begin{cases} R_{\text{TLS}}[\kappa] & \text{if } z_c[\kappa] = \hat{Q}_{\text{TLS}}[\kappa] \\ R_{\text{LS}}[\kappa] & \text{if } z_c[\kappa] = \hat{Q}_{\text{LS}}[\kappa] \end{cases} \tag{7.36}$$

Now, whenever a new measurement $z_c[\kappa] \in \left\{ \hat{Q}_{\text{TLS}}[\kappa], \hat{Q}_{\text{LS}}[\kappa] \right\}$ is received, the KF-based fused capacity estimate is obtained as

$$\begin{aligned} \hat{x}_c[\kappa|\kappa] = {} & \hat{x}_c[\kappa - 1|\kappa - 1] \\ & + \frac{P_c[k - 1|k - 1] + Q_c[\kappa - 1]}{P_c[\kappa - 1|\kappa - 1] + Q_c[\kappa - 1] + R_c[\kappa]} \left(z_c[\kappa] - \hat{x}_c[k - 1|k - 1] \right) \end{aligned} \tag{7.37}$$

where $\hat{x}_c[\kappa - 1|\kappa - 1]$ is the previous update of capacity estimate and $P_c[\kappa - 1|\kappa - 1]$ is the previous estimation error variance, which is updated as

$$P_c[\kappa|\kappa] = \frac{R_c[\kappa](P_c[\kappa - 1|\kappa - 1] + Q_c[\kappa - 1])}{P_c[\kappa - 1|\kappa - 1] + Q_c[\kappa - 1] + R_c[\kappa]} \tag{7.38}$$

Figure 7.7 shows a block diagram of the KF-based fusion approach for recursive estimation of capacity.

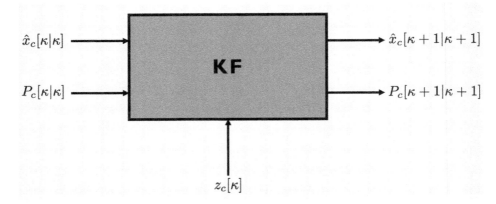

$\hat{x}_c[\kappa|\kappa]$

$P_c[\kappa|\kappa]$

KF

$\hat{x}_c[\kappa+1|\kappa+1]$

$P_c[\kappa+1|\kappa+1]$

$z_c[\kappa]$

Figure 7.7 Block diagram of the KF approach.

7.5 EXPERIMENTAL RESULTS

The proposed approach for estimating the capacity of a battery is demonstrated using data collected from a commercially available Li-ion battery. The model number of the battery is Samsung-30T INR21700. The battery is shown in Figure 7.8 and the features of the cell are summarized in Table 7.2. The tested battery is labeled C1212 and will be referred to using this label in the remainder of this chapter. The data from the battery is collected using the Arbin battery cycler (LBT21084, Arbin Instruments, USA). It has 16 independently controlled channels, each with a voltage range of 0 to 5V and a current range of ±10A. A single channel was used to collect data from the battery at room temperature.

A low current slow discharge-charge cycle is pursued to characterize the battery to model its OCV as a function of its SOC (Section 7.5.1). The true capacity of the battery is also determined by its OCV-SOC test. The characterization test is followed by a dynamic discharge-charge current profile (Section 7.5.2) during which the capacity of the battery is estimated 1 hour after every rest state in the profile.

7.5.1 OCV-SOC Characterization Test

In this section, the slow open-circuit voltage test that is performed to obtain the true total capacity and the OCV-SOC parameters of the battery is presented. Particularly, this section details how the data is collected for the total capacity and OCV parameter

Figure 7.8 Samsung-30T INR21700 Li-ion battery.

Table 7.2

Specifications of Li-ion Battery

Specification	Value (unit)
Nominal capacity	3000 mAh
Max. continuous discharge current	35A
Nominal voltage	3.6V
Height	70 mm
Diameter	21 mm
Weight	70g
Internal resistance	15 mΩ

estimation. Prior to the data collection for the OCV haracterization, the battery is fully charged using the CC-CV charging method. Algorithm 7.1 details how the battery is charged before the data collection for the OCV characterization begins. Here, the battery is declared fully charged when the charging current falls below the termination current, I_t. This termination current is selected to match the rate at which the OCV characterization data is collected. For example, the OCV characterization data is collected at C/30 rate; hence, the charge termination current is set to $I_t = C/30$.

Algorithm 7.1 CC-CV-Charge(I_t)

1: Measure terminal voltage v
2: **if** $v < v_n$ **then**
3:     CC-charge using $Q_r/4$
4:     Measure terminal voltage v
5:     **if** $v \geq \text{OCV}_{\max}$ **then**
6:         Go to line 9
7:     **end if**
8: **else**
9:     CV-charge at $v = \text{OCV}_{\max}$
10:     Measure current i
11:     **if** $i < I_t$ **then**
12:         Go to line 15
13:     **end if**
14: **end if**
15: Terminate Charging

Algorithm 7.2 describes the data collection for the OCV characterization experiment.

Algorithm 7.2 Slow-OCV-Test (N, T)

1: Set temperature: T
2: CC-CV-Charge(C/N)
3: 1-hour rest
4: CC-discharge (C/N)
    Sample data at 1/60 Hz
5: CC-charge (C/N)
    Sample data at 1/60 Hz
6: 1-hour rest

The terminal voltage and current are recorded from the battery during the OCV-SOC test. The data is processed to obtain the typical OCV-SOC curve represented by the combined+3 model in (7.39). The experimental setup used for testing the batteries can be found in Chapter 5.

The combined+3 OCV-SOC model is given by

$$V_\circ(s) = k_0 + \frac{k_1}{s} + \frac{k_2}{s^2} + \frac{k_3}{s^3} + \frac{k_4}{s^4}$$
$$+ k_5 s + k_6 \ln(s) + k_7 \ln(1-s) \tag{7.39}$$

where, it is assumed that the OCV parameters $k_0, k_1, \ldots k_7$ are obtained by linearly scaling $s \in [0,1]$ to lie between $[\epsilon, 1-\epsilon]$. For more details on the OCV parameter estimation approach, the reader is referred to Chapter 4.

Table 7.3 gives the values of the obtained OCV parameters for the battery cell C1212.

Table 7.3
Obtained OCV-SOC Parameters

	C1212
k_0	-9.931511
k_1	146.274963
k_2	-25.365256
k_3	2.845075
k_4	-0.139554
k_5	-113.475854
k_6	205.066794
k_7	-1.349234

The results of OCV-SOC modeling are shown in Figure 7.9. The true capacity of the battery is determined as the average of the discharge and charge capacities. The discharge and charge capacities are evaluated as the amount of Coulombs drawn out of the battery during the full discharge cycle and the amount of Coulombs supplied to the battery during the full charge cycle respectively. The evaluated true capacity of the battery C1212 is 2.9625 Ah.

7.5.2 Dynamic Discharge-Charge Profile

The battery is subjected to a dynamic discharge-charge current profile as shown in Figure 7.10. The profile mimics a driving cycle with rest states interspersed in discharge-charge cycles. The battery rest states are equivalent to when a vehicle stops. At these rest states,

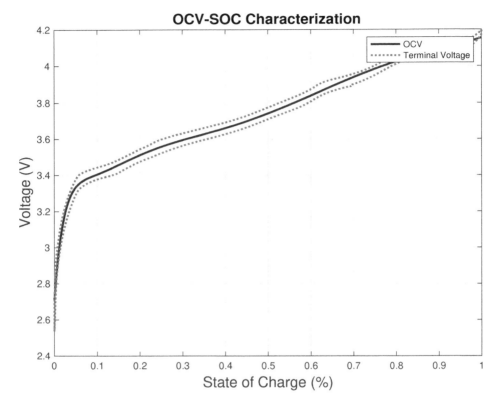

Figure 7.9 OCV-SOC characterization.

the battery is neither charging nor discharging (i.e., the battery current is zero). The proposed approach takes advantage of these rest stops to determine the capacity of the battery 1 hour after every rest is identified. In the experiment conducted, six rest states are identified (i.e., states at which the battery current is close to zero). One hour after every rest state is highlighted in dark gray and numbered, as shown in Figure 7.10.

7.5.3 Real-Time Capacity Estimation

As seen in Section 7.3, for six rest states, $N = 6$, 15 rest pairs are possible. The 15 possible rest pairs from the experiment are identified using Table 7.1. With the identified rest pairs, differential measurements of the form of (7.13) can be written. The following

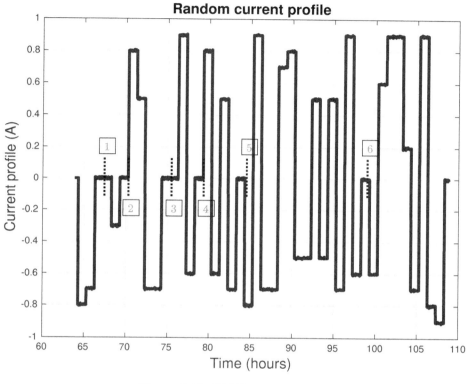

Figure 7.10 C1212 current profile.

observation model considering a batch of 15 measurements from 7.17 is written as

$$
\begin{bmatrix}
-0.1027 \\
-0.1333 \\
-0.0348 \\
-0.0356 \\
-0.2699 \\
-0.0306 \\
0.0679 \\
0.0670 \\
-0.1672 \\
0.0985 \\
0.0977 \\
-0.1366 \\
-0.00085 \\
-0.2351 \\
-0.2342
\end{bmatrix}
= Q^{-1}
\begin{bmatrix}
-0.2998 \\
-0.4018 \\
-0.1039 \\
-0.1054 \\
-0.8071 \\
-0.1020 \\
0.1959 \\
0.1943 \\
-0.5074 \\
0.2979 \\
0.2964 \\
-0.4053 \\
-0.0015 \\
-0.7032 \\
-0.7017
\end{bmatrix}
+ \tilde{\mathbf{w}}_s^\kappa
$$

A linear least-squares estimation method is pursued to estimate the capacity of the battery considering noisy measurements. The noise covariance matrix is of the form in (7.23) with $\sigma^2 = 0.001$. The estimated inverse of capacity by the least-squares estimator (7.21) is $Q^{-1} = 0.3340(Ah)^{-1}$. The transformation of the inverse capacity estimate following Taylor's series expansion in (B.9) is $Q = 3.0047$ Ah.

7.6 CONCLUSIONS

This chapter presented a real-time approach to estimating the total capacity of the battery. The proposed approach requires a minimum of two OCV measurements and the amount of Coulombs needed to go from one OCV measurement to the other. The advantage of this approach is that it does not require the complete discharge of the battery for capacity estimation. The OCV measurements can be corrupted by noise due to hysteresis and relaxation effects of the battery. To minimize errors, a least-squares approach was presented. The least squares approach uses a batch of OCV measurements and corresponding Coulomb computations to estimate the battery capacity such that the square error is minimized. When the Coulomb counting noise is significant, an alternate approach based on the total least-squares estimates was suggested. In order to reduce the variance in the estimates, a Kalman filter-based approach can be used to fuse estimates from multiple batches.

7.7 BIBLIOGRAPHICAL NOTES

Some elements of the capacity estimation approach detailed in this chapter are under active patent protection by [2], [3], and maybe others. For more details on the Kalman filter-based fusion approach for capacity estimation proposed in Section 7.4.3, the reader is directed to [8].

References

[1] B. Balasingam, G. V. Avvari, B. Pattipati, K. R. Pattipati, and Y. Bar-Shalom, "A robust approach to battery fuel gauging, part II: Real time capacity estimation," *Journal of Power Sources*, Vol. 269, pp. 949-961, 2014.

[2] B. Balasingam, B. French, Y. Bar-shalom, B. Pattipati, K. Pattipati, J. Meacham, T. Williams, G.V. Avvari, and T. S. Hwang, "Battery state of charge tracking, equivalent circuit selection and benchmarking," U.S. Patent Grant, No. US 10,664,562 B2, 2020.

[3] E. Barsoukov, D.R. Poole, and D.L. Freeman, "Circuit and method for measurement of battery capacity fade," U.S. Patent Grant, No. U.S. 6,892,148 B2, 2005.

[4] Texas Instruments, "1-4 series Li-ion battery pack manager supporting Turbo Mode 2.0," serial number BQ40Z50-R2, https://www.ti.com/product/BQ40Z50-R2, (accessed Dec. 2021).

[5] M. A. Roscher, O. Bohlen, J. Vetter, OCV hysteresis in li-ion batteries including two-phase transition materials, International Journal of Electrochemistry, 2011.

[6] R. Hamming, *Numerical methods for scientists and engineers*, Courier Dover Publications, 2012.

[7] I. Markovsky, and S. Van Huffel, "Overview of total least-squares methods," Signal Processing, Vol. 87, No. 10, pp. 2283-2302, 2017.

[8] B. Balasingam, and K. R. Pattipati, "On the identification of electrical equivalent circuit models based on noisy measurements," *IEEE Transactions on Instrumentation and Measurement*, Vol. 70, 2021.

[9] L. Crassidis, and Y. Cheng, "Error-covariance analysis of the total least square problem," *Journal of Guidance, Control, and Dynamics,* Vol. 37, No.4, pp. 1053-1063, 2014.

Chapter 8

Battery Fuel Gauging

8.1 INTRODUCTION

A battery fuel gauge (BFG) is the most important part of a BMS. The BFG estimates the SOC, SOH, time to shut down (TTS), and remaining useful life (RUL) of a battery based on three continuous measurements: the voltage across the battery terminals (terminal voltage), the current through the battery, and the temperature measured at various points on the surface of the battery. The BFG estimates are used in other important battery operations, such as optimized charging, charge balancing [1], and thermal balancing. Figure 8.1 shows a generic block diagram of a BFG.

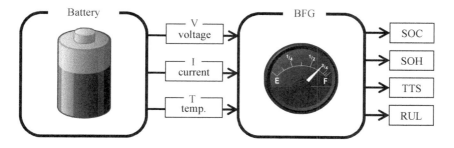

Figure 8.1 BFG. A BFG estimates the SOC, SOH, TTS, and RUL of a battery. Reproduced with permission from [2].

The challenge in BFG is that there is no direct way to measure the desired parameters, namely, SOC, TTS, SOH, and RUL. These quantities need to be estimated based on three direct measurements: voltage, current, and temperature. There is a non-linear relationship between the OCV and the SOC of the battery, and the parameters of this OCV-SOC curve need to be estimated for each battery chemistry through experimental and empirical approaches. The SOH of a battery has no direct relationship to any of the three direct measurements; consequently, the SOH must be computed for each individual battery using cumulative approaches. The internal impedance of the battery is needed to compute the available power; however, impedance changes due to battery age, temperature, and usage and needs to be continuously reestimated for accurate BFG output. Similarly, the battery capacity also changes with age, temperature, and usage and needs to be reestimated. Finally, measurement noise is a significant factor affecting the performance of a BFG.

8.1.1 State of Charge

The SOC is the remaining Coulombs in a battery as a ratio of the battery capacity, that is,

$$SOC = \frac{\text{Remaining charge}}{\text{Battery capacity}} \tag{8.1}$$

where $SOC \in [0, 1]$. It must be noted that battery capacity could vary based on many factors, such as temperature, age, and discharge rate.

Remark 8.1 The selection of battery capacity makes all the differences when computing the SOC. Many practical battery fuel gauge systems define SOC as a ratio of rated battery capacity that can be defined by the user (i.e., the battery engineer). The fuel gauge also allows the user to define the rate at which to measure the battery capacity. See Chapter 3 for examples of custom-defined capacity.

SOC is the most important BFG output that is required for important battery management operations such as charging, cell balancing, and safety. In an EV, the knowledge of SOC is required to continuously compute and update the remaining mileage of the vehicle. Accurate estimation of the remaining mileage is crucial to the safety and comfort of the occupants. Further, smart charging algorithms make use of the SOC estimate to reduce charging time without affecting the SOH of the battery. In electronic devices, such as smartphones, tablets, and laptop computers, SOC information is continuously updated and displayed. BMS in smartphones exploit historic SOC information of a particular phone to devise smart charging protocols that preserve battery health; for example, if an end user is found to be using only 40% of the battery on a daily basis, then the battery is only charged up to 70%. The SOC information is also used to make critical system decisions by the device operating system, such as forced hibernation when the SOC falls below a certain threshold (e.g., the smartphone shuts down when the SOC reaches 5%).

Remark 8.2 It is also possible that the device operating system maintains SOC in two different scales: one to display to the end user and the other for critical system decisions. For example, in most applications, the SOC display is rescaled in such a way that the end user sees the SOC values ranging from 0% to 100%. Here, what the user sees is a scaled version of the true SOC as defined by the system engineer.

Remark 8.3 The user of a BMS is the system engineer. For example, in designing a smartphone, the BMS engineer is responsible to select a suitable BMS (from such manufacturers as Texas Instruments and Analog Devices) and configure it such that the BMS communicates with the operating system of the device. The end user is the one who uses the smartphone.

The SOC cannot be directly measured on a battery; rather, it must be computed based on other direct measurements: voltage, current, and temperature. SOC estimation remains one of the active research topics in battery research. Two prominent approaches to SOC estimation are Coulomb counting and voltage-based lookup; both of these approaches suffer from numerous sources of error; a fusion-based approach is developed in order to combine the benefits of both the voltage-based and current-based approaches. The fusion-based approach is often implemented using the extended Kalman filter. The extended Kalman filter-based approach to SOC estimation has received significant attention in the literature; however, model uncertainty remains a significant problem.

8.1.2   Time to Shut Down

Rechargeable batteries in general are extremely sensitive to voltage breaches. Subjecting a Li-ion battery to over-voltage (or over-charging) will trigger thermal runaway is an irreversible process causing the battery to melt down and catch fire. Most rechargeable batteries also suffer from low-voltage breach. For example, when the terminal voltage of the battery drops below a certain threshold (this may happen due to over-discharging), the battery might be permanently damaged; consequently, the damaged cell might become a short-circuit and cause damage to adjacent battery cells and eventually to the entire battery pack. Hence, it is important for a BMS to compute the TTS during charging as well as during discharging so that battery can be electronically isolated from the charger/load to prevent damage to the battery pack. In addition to safety over-discharge protection (which is typically set much lower, e.g., 2.7–3V for a single cell

Li-ion battery), there are system limitations for minimal discharge voltage; that is, the battery may not be able to power the system at such low voltages.

Figure 8.2 describes the TTS requirement while charging or discharging a battery under two different scenarios. In one scenario, indicated in purple, the magnitude of the current is high and causes a voltage drop of approximately 0.1V; that is, the terminal voltage will be 0.1V above the OCV during charging and it will be 0.1V below the OCV during discharging. At this rate, the battery needs to be shut down when the SOC is just below 80% during charging and when the SOC is about 15% during discharging. Here, the safe voltage range is assumed to be between 3.5V (discharge threshold) and 4.2V (charge threshold). In the second scenario, the magnitude of the current is low causing a voltage drop of about 0.03V; at this rate, the battery can be charged to about 95% SOC and discharged to about 5% under the same voltage thresholds. For the end user, it would seem like the battery would not charge beyond 95% and that it dies at 5%. Reporting the SOC based on the discharge capacity at that (average) current will avoid this confusion to the end user, as described in Remark 8.2.

Figure 8.2 also shows why a battery cannot be fully charged (a battery is full when its SOC reaches 100%) using a high current. Indeed, any current more than zero will not fully charge the battery; this is due to the strict adherence to the voltage protection mechanism that is an integral part of modern battery cells. In order to fully charge a battery, the current must be reduced; this can be achieved by a constant voltage (CV) charging topology. During CV charging, the current gradually reduces towards zero; when the current falls below a certain threshold, below the C/20 rate in most chargers, the charger is stopped. Typical fast charging algorithms employ a constant (high) current at the start and then switch to CV charging, that is, fast chargers employ a constant-current constant-voltage (CC-CV) charging strategy. Finally, Figure 8.2 also explains why the discharge capacity of a battery decreases with increasing discharge current.

8.1.3 State of Health

Two major indicators of the SOH of a battery are power fade and capacity fade. Power fade indicates the percentage increase of the internal impedance of the battery and CF indicates the percentage decrease in the battery capacity. Both PF and CF need to be computed at a fixed temperature.

Due to solid electrolyte interface (SEI) formation (and growth) and other internal chemical reactions within a battery, its impedance increases over time. When the impedance of the battery increases, the output power from it decreases; this phenomenon

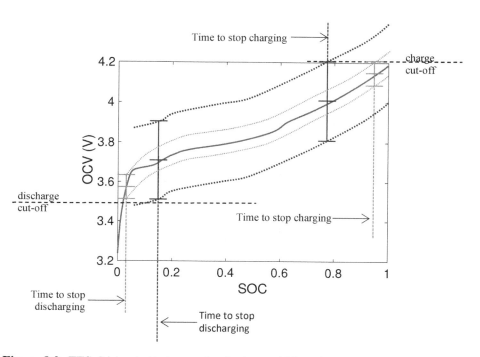

Figure 8.2 TTS. Li-ion batteries need to be kept within strict upper and lower voltage limits. During charging and discharging, the terminal voltage cannot exceed these voltage limits. The higher the amplitude of the charging/discharging current, the shorter the TTS is. Reproduced with permission from [2].

is known as the power fade.

$$\text{PF}(k) = \left(1 - \frac{P(k)}{P(0)}\right) 100 \ \% \tag{8.2}$$

where the available power at time k is defined as

$$P(k) = v\left(\frac{V_o(s) - v}{R_0(k)}\right) \tag{8.3}$$

$R_0(k)$ is the ohmic resistance at time k, $V_o(s)$ is the OCV of the battery at a certain reference SOC s, and v is the measured voltage across battery terminals. Here, the time index k indicates the elapse of life-cycle event, such as time and charge-discharge cycle; it is assumed that $k = 0$ denotes the very initial cycle (e.g., brand-new battery). The above definition of PF is only in terms of the series ohmic resistance R_0. In reality, the battery behavior is represented using higher-order electrical equivalent circuit models.

Remark 8.4 The ohmic resistance of a battery increases by several hundred percent at low temperatures. Hence, PF must always be computed at the same temperature.

Remark 8.5 The definition of power in (8.3) tells us that the power may differ at different SOC values. Hence, it is also important to compute PF at a certain SOC.

As the battery ages, the amount of lithium ions decreases due to various chemical reactions within the battery cell. The likelihood of self-discharge also increases with aging; this reduces the battery capacity. The capacity fade is formally defined as

$$\text{CF}(k) = \left(1 - \frac{Q(k)}{Q(0)}\right) 100\ \% \tag{8.4}$$

where $Q(0)$ is the initial capacity of the battery and $Q(k)$ is the capacity at time k; similar to before, time k could be either a cycle number or a calendar time unit.

Remark 8.6 Even though battery capacity does not fluctuate as wide as the Ohmic resistance, capacity does change with temperature. Hence, CF must always be computed at the same temperature.

In practice, the exact values of the resistance $R_0(k)$ and capacity $Q(k)$ of the battery at time k are not known; they need to be estimated. Hence, the PF and CF equations need to be rewritten as

$$\text{PF}(k) = \left(1 - \frac{\hat{P}(k)}{P(0)}\right) 100\ \% \tag{8.5}$$

$$\text{CF}(k) = \left(1 - \frac{\hat{Q}(k)}{Q(0)}\right) 100\ \% \tag{8.6}$$

where $\hat{Q}(k)$ is the estimated values of the capacity and $\hat{P}(k)$ is computed based on the estimated value of the resistance R_0, that is, by replacing R_0 with its estimated value $\hat{R}_0$ in (8.3). That is, in order to compute its final outputs (SOC, SOH, TTS, and RUL), a BFG needs to estimate the capacity and the impedance of the battery. The different approaches to estimating these parameters are the subject of very active ongoing research.

There is always an uncertainty in an estimated parameter; the associated uncertainty of an estimated parameter is usually denoted by the variance of the estimator; that is,

$$\sigma^2_{R_0(k)} = E\left[\left(\hat{R}_0(k) - E[\hat{R}_0(k)]\right)^2\right] \qquad (8.7)$$

$$\sigma^2_{Q(k)} = E\left[\left(\hat{Q}(k) - E[\hat{Q}(k)]\right)^2\right] \qquad (8.8)$$

where $\sigma^2_{R_0(k)}$ denotes the variance of estimating $R_0(k)$ and $\sigma^2_{Q(k)}$ denotes the variance of estimating $Q(k)$. Usually, as the battery ages, the uncertainty in estimating these two parameters also increases.

Figure 8.3(a) shows a generic plot of the PF (i.e., the increase of resistance R_0 over time) in a battery. Possible estimated values of $R_0(k)$ by a hypothetical BFG are shown along the true value at discrete time instances k. The uncertainty of $R_0(k)$ estimation is indicated in dashed lines with ellipses. As previously mentioned, the uncertainty of estimation is indicated to be increasing with battery age.

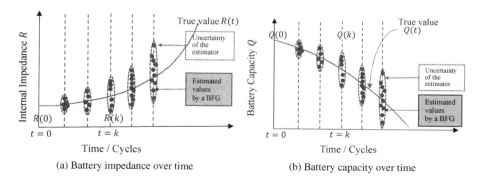

(a) Battery impedance over time   (b) Battery capacity over time

Figure 8.3 (a, b) SOH of a battery. Two major indicators of the SOH are PF and CF, which are computed in terms of the estimated resistance and capacity, respectively, of the battery. Estimated values are always accompanied by uncertainty; the better the estimator, the lower the uncertainty. This plot is not based on experimental data; it only serves to illustrate the idea that as the battery gets older, the uncertainty around parameter estimation is likely to increase. Reproduced with permission from [2].

Similar to the $R_0(k)$ estimation in Figure 8.3(a), the true value of capacity and its possible estimated values by a BFG is shown in Figure 8.3(b). Similar to before, the uncertainty of the estimation is shown to increase over time.

Having defined PF and CF, the SOH of the battery is a means to unify them and to express the state of health using a single quantity. In applications where the output power is more critical, the SOH could be defined as

$$\mathrm{SOH} = 100 - \mathrm{PF} \tag{8.9}$$

However, in applications where running time is more critical than the peak output power, the SOH could be defined as

$$\mathrm{SOH} = 100 - \mathrm{CF} \tag{8.10}$$

In applications where both power and capacity are critical, SOH could be defined through various design criteria; for instance, a stricter SOH measure is

$$\mathrm{SOH} = 100 - \max\left\{\mathrm{PF}, \mathrm{CF}\right\} \tag{8.11}$$

and a less strict SOH measure is

$$\mathrm{SOH} = 100 - \min\left\{\mathrm{PF}, \mathrm{CF}\right\} \tag{8.12}$$

where $\mathrm{SOH} = 100\%$ implies a brand-new battery and $\mathrm{SOH} = 0\%$ implies a completely used up battery.

It must be noted that all SOH values are computed based on the assumption that the initial values (i.e., $P(0)$ and $Q(0)$) are known. Further, the initial values $P(0), Q(0)$ and the present values $P(k), Q(k)$ should be computed at the same temperature and SOC in order to reduce variability from those factors. The state-of-the-art BFG can only compute the SOH of a known battery. There are no ways to compute the SOH when the initial values of $P(0)$ and $Q(0)$ are not known.

8.1.4 Remaining Useful Life

As the batteries age, they need to be replaced in critical applications. For example, an EV with an aged battery may not be able to achieve its target speed on highways (due to PF); this is a serious safety issue. Hence, it is important to predict when a battery pack will reach a specific SOH threshold so that battery replacement can be planned ahead of time.

Based on any of the definitions of SOH in (8.9) through (8.12), the SOH of a brand-new battery is 100%. As the battery ages, the SOH reduces; many applications consider the battery dead when the SOH reaches 80%. This practice is different from the practice used for the SOC display (see Remark 8.2). The RUL of a battery is defined as the expected time, in terms of the number of charge-discharge cycles, a battery pack will take until it reaches a certain SOH threshold that is predefined by the user.

Figure 8.4 describes the phenomenon of the RUL in terms of the predicted number of cycles until the SOH drops to a specific threshold. Similar to the case of PF and CF estimation, one can expect the uncertainty of the SOH to increase with battery aging. For example, in the case of CF, the estimated capacity becomes less reliable due to age-related changes in the OCV-SOC curve. Similarly, PF estimation is met with increasing uncertainty due to chemical changes within the battery, such as solid electrolyte interface formation [1]. Figure 8.4 shows such prediction uncertainties of two different BFGs; the uncertainty of one BFG has a lower uncertainty compared to the uncertainty of the other BFG. Figure 8.4 demonstrates that having a good BFG is analogous to extending the life of a battery.

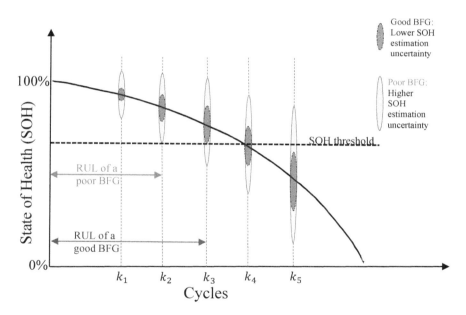

Figure 8.4 RUL: same battery, two different mileages. RUL is the amount of time, in terms of the number of cycles, it takes until the SOH of the battery drops to a predetermined threshold. The uncertainties of two different BFGs are compared to demonstrate the importance of BFG in extending the usable life of a battery. Reproduced with permission from [2].

Battery aging is one of the limiting factors of wider EV adaptation. Currently, replacing an EV battery is costly and time-consuming. Therefore, significant research is being done to determine newer and better methods to increase the RUL of a battery. Some important avenues of ongoing research are:

- Search of new battery chemistries: There is ongoing research about finding new battery chemistries that can offer desired qualities in a rechargeable battery: fast charging without the risk of thermal runaway and reduced SOH. For example, solid-state batteries have been shown to possess such qualities; however, solid-state batteries remain expensive and more research and investment are needed before they can be widely adopted.

- Optimal charging strategies: Fast charging causes battery health deterioration. Consider an EV battery that has a capacity of 300 Ah. In order to charge this battery in 15 minutes, the charging current must be 1200A; such high currents, in addition to causing heat loss, cause lithium plating and result in reduced battery capacity. It is believed that better charging strategies can reduce some of the detrimental effects of fast charging. Pulse charging is one such strategy studied by researchers for faster charging of batteries without severely affecting the SOH. Future EV fleets could also use constant connectivity for on-demand, strategic charging instead of fast charging.

- Improved BFG algorithms: The research community is actively studying and improving approaches to estimate BFG outputs. As suggested in Figure 8.3, more accurate BFG algorithms can help to improve the RUL. Also, accurate SOC and TTS information is crucial for the success of optimal charging algorithms.

8.2 SOC ESTIMATION: COULOMB COUNTING APPROACH

The SOC of a battery can be updated through the following Coulomb counting equation

$$s(t) = s(0) + \frac{\eta}{3600 \mathrm{C}_{\mathrm{batt}}} \int_0^t i(t) dt \qquad (8.13)$$

where η is the Coulomb counting efficiency defined as follows:

$$\eta = \begin{cases} \eta_c & i(t) > 0 \\ \eta_d & i(t) < 0, \end{cases} \qquad (8.14)$$

t denotes the unit of time in seconds, $i(t)$ is the current in amperes (A) through the battery at time t, $s(0)$ denotes the SOC at time instant $t = 0$, $s(t)$ shows the SOC at time interval t, and C_{batt} denotes the battery capacity in ampere-hours (Ah).

A discretized version of the Coulomb counting equation is given as follows:

$$s(k) = s(k-1) + \frac{\eta}{3600C_{\text{batt}}} \int_{t(k-1)}^{t(k)} i(\tau)d\tau \tag{8.15}$$

where $s(k)$ denotes the SOC at time instant $t(k)$, and $i(\tau)$ shows the measured current at time instant τ. The approximation is done using the rectangular (backward difference) method for the integration in (8.15) as follows:

$$\int_{t(k-1)}^{t(k)} i(\tau)d\tau \approx \Delta_k i(t(k)) = \Delta_k i(k) \tag{8.16}$$

where $\Delta_k = t(k) - t(k-1)$, the Coulomb counting equation, is further simplified as follows:

$$s(k) = s(k-1) + \frac{\eta \Delta_k i(k)}{3600C_{\text{batt}}} \tag{8.17}$$

The Coulomb counting equation in (8.17) suffers from the following sources of errors:

1. Measurement error in the current $i(k)$.

2. Error because of the approximation used for the integration in (8.16).

3. Unreliability in the battery capacity knowledge C_{batt}.

4. Uncertainty in the knowledge of the Coulomb counting efficiency η.

5. Measurement error in sampling time Δ.

Next, detailed mathematical discussions about these four types of errors are provided.

8.3   SOC ESTIMATION: AN OCV-BASED APPROACH

The voltage-based SOC estimation approach is basically a table look-up method. Battery terminals' voltage is matched with the SOC according to the OCV-SOC characterization curve.

In more generic terms, we have the following:

$$\text{voltage-measurement} = \underbrace{f(\text{SOC})}_{\text{OCV-SOC model}} + \underbrace{g(\text{parameters, current})}_{\text{voltage drop}} \qquad (8.18)$$

where the function $f(\cdot)$ refers to the open-circuit voltage model that relates the OCV of the battery to SOC and $g(\cdot)$ accounts for the voltage drop within the battery cell, due to hysteresis and relaxation effects. Challenges in voltage-based SOC estimation arise due to the fact that the functions $f(\cdot)$ and $g(\cdot)$ are usually nonlinear and that there is a great amount of uncertainty as to what those functions might be. For instance, the parameters can be modeled through electrical ECM or electrochemical models, each of which can result in numerous reduced-order approximations. The voltage-based approach suffers from the following three types of errors:

1. OCV-SOC modeling error: The OCV-SOC relationship of a battery can be approximated through various models: linear model, polynomial model, and combined models are a few examples. Reducing the OCV-SOC modeling error is an ongoing research problem—in [3], a new modeling approach was reported that resulted in the "worst case modeling error" of about 10 mV. It must be mentioned that the OCV modeling error is not identical in all voltage regions of the battery.

2. Voltage-drop modeling error: Voltage-drop models account for the hysteresis and relaxation effects in the battery. Various approximations were proposed in the literature in order to represent these effects.

3. Voltage measurement error: Every voltage measurement system comes with errors; this translates into a SOC estimation error.

In order to reduce the effect of uncertainties in voltage-based SOC estimation, it is often suggested to rest the battery before taking the voltage measurement for SOC, when the current is zero for a sufficient time the voltage drop also approaches zero. However, all the other sources of errors mentioned above (OCV-SOC modeling error, hysteresis, and voltage measurement error) cannot be eliminated by resting the battery.

8.4 SOC ESTIMATION: FUSION APPROACH

The traditional Coulomb counting model (8.17) suffers from several limitations due to uncertainties in initial SOC, current measurement, current integration approximation, battery capacity, charging/discharging efficiencies, and sampling time. It was shown in [4] that the traditional Coulomb counting equation can be corrected as follows:

$$s(k) = s(k-1) + \frac{\Delta_k z_i(k)}{3600Q} + n_s(k) \tag{8.19}$$

where $n_s(k)$ accounts for the error or noise due to all the uncertainties and $z_i(k)$ denotes the measured current that is modeled as

$$z_i(k) = i(k) + n_i(k) \tag{8.20}$$

The current measurement noise $n_i(k)$ is assumed to be zero-mean white with standard deviation σ_i.

8.4.1 Measurement Model

Figure 8.5(a) shows the equivalent circuit model of a battery. The measured battery terminal voltage $v(k)$ is

$$\begin{aligned} z_v(k &= v(k) + n_v(k) \\ &= V_o(s(k)) + v_d(k) + n_v(k) \end{aligned} \tag{8.21}$$

where $V_o(s(k))$ is the open-circuit voltage, which is a function of the battery SOC $s(k)$, $v_d(k)$ is the voltage drop across the internal components of the battery, and $n_v(k)$ is the measurement noise which is assumed zero-mean with standard deviation σ_v.

There are several functions to approximate the OCV-SOC relationship. Here, the following combined+3 model is adopted:

$$\begin{aligned} V_o(s(k)) =&k_0 + \frac{k_1}{s(k)} + \frac{k_2}{s^2(k)} + \frac{k_3}{s^3(k)} + \frac{k_4}{s^4(k)} \\ &+ k_5 s(k) + k_6 \ln(s(k)) + k_7 \ln(1 - s(k)) \end{aligned} \tag{8.22}$$

The voltage drop, v_d, is due to hysteresis and relaxation effects in a battery.

Several reduced-order equivalent-circuit models were proposed to represent the voltage drop. Figure 8.5(b) shows an approximation consisting of internal resistance and two RC circuits. Some well-known voltage-drop approximations are:

$$v_d(k) = \begin{cases} i(k)R_0 & \text{R-int Model} \\ i(k)R_0 + i_1(k)R_1 & \text{RC Model} \\ i(k)R_0 + i_1(k)R_1 + i_2(k)R_2 & \text{2RC Model} \\ i(k)R_0 + i_1(k)R_1 + i_2(k)R_2 + h(k) & \text{ESC Model} \end{cases} \tag{8.23}$$

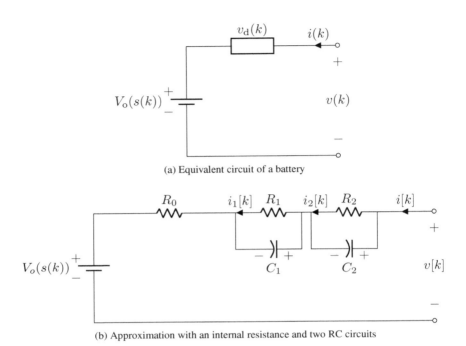

(a) Equivalent circuit of a battery

(b) Approximation with an internal resistance and two RC circuits

Figure 8.5 (a, b) Battery ECM approximation.

where ESC denotes the enhanced self-correcting model and $h(k)$ denotes the hysteresis voltage (see Chapter 3 for more on the hysteresis model). The parameters $R_0, R_1, R_2, i_1(k), i_2(k)$, and $h(k)$ are unknown quantities and need to be estimated. Estimating these parameters with high certainty remains an active research problem.

8.4.2 Scaling

It can be noticed that the SOC $s(k)$ in (8.22) can take any values between 0 and 1. This will cause numerical instability when $s(k) \rightarrow 0$ and when $s(k) \rightarrow 1$. To tackle this problem, the linear scaling approach proposed in [3] can be exploited. Using linear scaling, the scaled SOC can be written as

$$s'(k) = s(k)(1 - 2\epsilon) + \epsilon \tag{8.24}$$

where $0 < \epsilon < 0.5$ and $s'(k) \in [\epsilon, 1 - \epsilon]$.

The Coulomb counting model (8.19) can be rewritten for scaled SOC as

$$\frac{s'(k)}{1-2\epsilon} = \frac{s'(k-1)}{1-2\epsilon} + \frac{\Delta_k z_{\mathrm{i}}(k)}{3600Q} + n_{\mathrm{s}}(k)$$

$$= s'(k-1) + \underbrace{\frac{(1-2\epsilon)\Delta_k z_{\mathrm{i}}(k)}{3600Q}}_{G(k)} - \underbrace{(1-2\epsilon)n_{\mathrm{s}}(k)}_{v_{\mathrm{s}}(k)}$$

By denoting the scaled SOC at time k as

$$x_{\mathrm{s}}(k) \triangleq s'(k) \tag{8.25}$$

the process model for SOC estimation becomes

$$x_s(k) = x_s(k-1) + G(k)z_{\mathrm{i}}(k) + v_{\mathrm{s}}(k) \tag{8.26}$$

where

$$G(k) = \frac{(1-2\epsilon)\Delta_k}{3,600Q} \tag{8.27}$$

and the process noise $v_{\mathrm{s}}(k)$ can be shown to have zero mean with standard deviation $(1-2\epsilon)\sigma_s$.

8.4.3 Extended Kalman Filter for SOC Tracking

A generic process model for SOC estimation will be derived in this section as follows

$$\mathbf{x}(k+1) = \mathbf{F}(k)\mathbf{x}(k) + \mathbf{u}(k)z_{\mathrm{i}}(k) + \mathbf{v}(k) \tag{8.28}$$

where $z_{\mathrm{i}}(k)$ is the measured current. The state vector $\mathbf{x}(k)$, the state transition matrix $\mathbf{F}(k)$, the input $\mathbf{u}(k)$, and the process noise $\mathbf{v}(k)$ can vary based on the assumptions made for the voltage drop model. These variables will be defined for a particular model assumption. A generic process noise covariance is defined as

$$\mathbf{Q} = E\left\{\mathbf{v}(k)\mathbf{v}(k)^T\right\} \tag{8.29}$$

Based on (8.21), let us write a generic measurement model as follows

$$\mathbf{z}(k) = h(\mathbf{x}(k)) + \mathbf{w}(k) \tag{8.30}$$

Correspondingly, the generic measurement noise covariance is defined as

$$\mathbf{R} = E\left\{\mathbf{w}(k)\mathbf{w}(k)^T\right\} \tag{8.31}$$

Next, two different cases are elaborated to satisfy the above state-space models under varying assumptions on the voltage drop model.

8.4.3.1   Case 1: R-int Model

For the R-int model, the voltage drop (8.23) reduces to $v_\mathrm{d}(k) = i(k)R_0$ and the measurement model can be written as

$$
\begin{aligned}
z_\mathrm{v}(k) &= V_\mathrm{o}(x_\mathrm{s}(k)) + i(k)R_0 + n_\mathrm{v}(k) \\
&= V_\mathrm{o}(x_\mathrm{s}(k)) + (z_i(k) - n_i(k))R_0 + n_\mathrm{v}(k) \\
&= V_\mathrm{o}(x_\mathrm{s}(k)) + z_\mathrm{i}(k)R_0 + w(k)
\end{aligned} \tag{8.32}
$$

where the voltage measurement noise $w(k)$ is zero-mean Gaussian with standard deviation σ_w given by

$$\sigma_w^2 = \sigma_\mathrm{v}^2 + \sigma_\mathrm{i}^2 R_0^2 \tag{8.33}$$

The specific process model for the R-int assumption can be obtained through the following replacements in (8.28)

$$\mathbf{x}(k) \leftarrow x_s(k), \ \mathbf{F}(k) \leftarrow 1, \ \mathbf{u}(k) \leftarrow G(k), \ \mathbf{v}(k) \leftarrow v_\mathrm{s}(k) \tag{8.34}$$

where the process noise covariance becomes a scalar, that is, $\mathbf{Q} \leftarrow Q$ and

$$Q = E\left(v(k)^2\right) = (1 - 2\epsilon)^2 \sigma_s^2 \tag{8.35}$$

Similarly, the specific measurement model under the R-int assumption is obtained through the following replacement

$$h(\mathbf{x}(k)) \leftarrow V_\mathrm{o}(x_\mathrm{s}(k)) + z_\mathrm{i}(k)R_0, \ \mathbf{w}(k) \leftarrow w(k) \tag{8.36}$$

8.4.3.2   Case 2: 2RC Model

For 2RC model, the voltage drop (8.23) reduces to $v_\mathrm{d}(k) = i(k)R_0 + i_1(k)R_1 + i_2(k)R_2$ and the measurement model can be written as

$$z_\mathrm{v}(k) = V_\mathrm{o}(x_\mathrm{s}(k)) + i(k)R_0 + i_1(k)R_1 + i_2(k)R_2 + n_\mathrm{v}(k) \tag{8.37}$$

Here, the currents through R_1 and R_2 are computed recursive as follows:

$$i_1(k) = \alpha_1 i_1(k-1) + (1-\alpha_1)i(k)$$
$$i_2(k) = \alpha_2 i_2(k-1) + (1-\alpha_2)i(k)$$
(8.38)

where the quantities $i_1(k)$ and $i_2(k)$ need to be estimated. For this, let us denote these two quantities as follows:

$$x_1(k) \triangleq i_1(k)$$
$$x_2(k) \triangleq i_2(k)$$
(8.39)

By considering measured value for the current $i(k)$, the recursions in (8.38) can be rewritten as

$$x_1(k) = \alpha_1 x_1(k-1) + (1-\alpha_1)z_i(k) + v_1(k)$$
(8.40)
$$x_2(k) = \alpha_2 x_2(k-1) + (1-\alpha_2)z_i(k) + v_2(k)$$
(8.41)

where v_1 and v_1 are zero-mean Gaussian noise with variances given by

$$\sigma_1^2 = \sigma_i^2(1-\alpha_1)^2$$
(8.42)
$$\sigma_2^2 = \sigma_i^2(1-\alpha_2)^2$$
(8.43)

Now, similar to (8.32), the measurement model (8.37) can be rewritten in terms of the state variables as follows:

$$z_v(k) = V_o(x_s(k)) + z_i(k)R_0 + x_1(k)R_1 + x_2(k)R_2 + w(k)$$
(8.44)

The process model for the 2RC model assumption can be obtained through the following replacements:

$$\mathbf{x}(k) \leftarrow \begin{bmatrix} x_s(k) \\ x_1(k) \\ x_2(k) \end{bmatrix}, \quad \mathbf{F}(k) \leftarrow \begin{bmatrix} 1 & 0 & 0 \\ 0 & \alpha_1 & 0 \\ 0 & 0 & \alpha_2 \end{bmatrix}$$
$$\mathbf{u}(k) \leftarrow \begin{bmatrix} G(k) \\ 1-\alpha_1 \\ 1-\alpha_2 \end{bmatrix}, \quad \mathbf{v}(k) \leftarrow \begin{bmatrix} v_s(k) \\ v_1(k) \\ v_2(k) \end{bmatrix}$$
(8.45)

where the process noise covariance is

$$\mathbf{Q} = E\left(\mathbf{v}(k)\mathbf{v}(k)^T\right) \tag{8.46}$$

$$\mathbf{Q} = \begin{bmatrix} (1-2\epsilon)^2\sigma_s^2 & 0 & 0 \\ 0 & \sigma_i^2(1-\alpha_1)^2 & 0 \\ 0 & 0 & \sigma_i^2(1-\alpha_2)^2 \end{bmatrix} \tag{8.47}$$

Similarly, the specific measurement model under the 2RC assumption is obtained through the following replacements

$$\begin{aligned} h(\mathbf{x}(k)) &\leftarrow V_\circ(x_s(k)) + z_i(k)R_0 + x_1(k)R_1 + x_2(k)R_2 \\ \mathbf{w}(k) &\leftarrow w(k) \end{aligned} \tag{8.48}$$

So far, two specific cases for the process model (8.28) and the measurement model (8.30) were derived in Sections 8.4.3.1 and 8.4.3.2, respectively. In both cases, the process model is linear and the measurement model is nonlinear. In order to implement the EKF algorithm, the following linearization will be needed:

$$\mathbf{H}(k+1) = \left.\frac{\partial h(\mathbf{x}(k))}{\partial \mathbf{x}(k)}\right|_{\mathbf{x}(k)=\hat{\mathbf{x}}(k+1|k)} \tag{8.49}$$

where $\mathbf{H}(k+1)$ is the linearized observation model and $\hat{\mathbf{x}}(k+1|k)$ is the predicted state vector. Depending on the model assumption, the linearization will vary:

$$\mathbf{H}(k+1) = \begin{cases} V_\circ(\hat{x}_s(k+1|k))' & \text{R-int} \\ \begin{bmatrix} V_\circ(\hat{x}_s(k+1|k))' & R_1 & R_2 \end{bmatrix} & \text{2RC} \end{cases} \tag{8.50}$$

where $V_\circ(\hat{x}_s(k+1|k))'$ is the derivative of (8.22) with respect to the state of charge.

Having defined the process and measurement models in (8.28) and (8.30), respectively, the recursive SOC estimation problem can be formally stated as follows. Given the voltage and current measurements, $z_v(k)$ and $z_i(k)$, respectively, and the state estimates $\hat{\mathbf{x}}(k|k)$ and corresponding estimation error covariance $\mathbf{P}(k|k)$ at time k, compute the updated state estimate and the corresponding estimation error covariance $\hat{\mathbf{x}}(k+1|k)+1$ and $\mathbf{P}(k+1|k+1)$, respectively.

The EKF algorithm (see Chapter 2) works by taking as an input the previous state $\hat{\mathbf{x}}(k|k)$, previous covariance $\mathbf{P}(k|k)$, current measurement $z_i(k+1)$, and voltage measurement $z_v(k+1)$ and outputs the state $\hat{\mathbf{x}}(k+1|k+1)$ and covariance estimate $\mathbf{P}(k+1|k+1)$. In the process, it calculates the state prediction $\hat{\mathbf{x}}(k+1|k)$, state prediction variance $\mathbf{P}(k+1|k)$, measurement prediction $\hat{z}(k+1|k)$, the innovation variance $S(k+1|k)$, innovation $\nu(k+1)$, and filter gain $W(k+1)$.

8.5 FILTER CONSISTENCY TESTING APPROACHES

Typical battery SOC estimation algorithms have to run continuously. This requires continuous performance monitoring in order to ensure that the EKF produces consistent estimates. In this section, two metrics are presented for consistency checking of the EKF algorithm.

8.5.1 Normalized Innovation Squared

The normalized innovation squared (NIS) is defined as [5]

$$\epsilon_\nu(k) = \frac{\nu(k)^2}{S(k)} \tag{8.51}$$

where $\nu(k)$ is the filter innovation and $S(k)$ is the variance of the innovation at time k. Considering the window length of T_w consecutive NIS values, the average NIS is given as

$$\bar{\epsilon}_\nu(k) = \frac{1}{T_w} \sum_{i=1}^{T_w} \epsilon_\nu^i(k) \tag{8.52}$$

It can be shown that $T_w \bar{\epsilon}_\nu$ is chi-square-distributed with T_w degrees of freedom. Then, for a certain level of confidence $(1 - \alpha)$, one can write

$$r_1 \le T_w \bar{\epsilon}_\nu(k) \le r_2 \tag{8.53}$$
$$p\left(r_1 \le T_w \bar{\epsilon}_\nu(k) \le r_2\right) = 1 - \alpha$$
$$\frac{r_1}{T_w} \le \bar{\epsilon}_\nu(k) \le \frac{r_2}{T_w}$$

where $[r_1, r_2]$ is the two sided chi-square confidence interval with T_w degrees of freedom for the confidence level α. That is, r_1 and r_2 are found such that a chi-square probability density function with T_w degrees of freedoms has $1 - \alpha$ as the area of the PDF between r_1 and r_2. Here, $T_w = 300$ is selected as the window length. For 99% confidence level (i.e., for $\alpha = 0.01$, and $T_w = 300$), it can be shown that $r_1 = 240.66$ and $r_2 = 366.84$, that is,

$$\frac{240.66}{T_w} \le \bar{\epsilon}_\nu(k) \le \frac{366.84}{T_w} \tag{8.54}$$
$$0.8022 \le \bar{\epsilon}_\nu(k) \le 1.2228$$

8.5.2   Zero-Mean Test of Innovations

For a consistent filter, the filter innovations $\nu(k)$ must be zero-mean. By selecting T_w consecutive filter innovations, the student's t-test is performed to test whether the mean of innovation is zero. Here, the one-sample t-test is employed to test the null hypothesis that the innovation is zero means against the alternative hypothesis (i.e., the mean is nonzero).

Consider a window of length T_w and window number L; the innovations for that window L are given as

$$\boldsymbol{\nu}_L = \begin{bmatrix} \nu((L-1)T_w + 1) \\ \nu((L-1)T_w + 2) \\ \vdots \\ \nu(LT_w) \end{bmatrix} \tag{8.55}$$

For the window $\boldsymbol{\nu}_L$ of length T_w, let us define the average and standard deviation as

$$\bar{\nu}_L = \frac{1}{T_w} \sum_{i=(L-1)T_w+1}^{LT_w} \boldsymbol{\nu}_L(i) \tag{8.56}$$

$$s_L = \sqrt{\frac{(\sum_{i=(L-1)T_w+1}^{LT_w}(\boldsymbol{\nu}_L(i) - \bar{\nu}_L)^2}{T_w - 1}} \tag{8.57}$$

Here, $\bar{\nu}_L$ is distributed according to the student's t-distribution with T_w degrees of freedom. Now, the t-statistic is written as

$$t(\boldsymbol{\nu}_L) = \frac{\bar{\nu}_L}{\frac{s_L}{\sqrt{T_w}}} \tag{8.58}$$

For the null hypothesis, that the filter innovations are zero-mean, to be true, with confidence $1 - \alpha$, the two-sided interval for $(T_w - 1)$ degrees of freedom is given as

$$-t_1 \le t(\boldsymbol{\nu}_L) \le t_2 \tag{8.59}$$

For $T_w = 300$ and $\alpha = 0.01$, the two-sided limits for the mean of innovation to be zero are

$$-2.5924 \le t(\boldsymbol{\nu}_L) \le 2.5924 \tag{8.60}$$

Remark 8.7 For $T_w = 300$ number of samples, the t-distribution can be approximated by a standard Gaussian distribution. The proposed consistency test is applicable to samples T_w of all sizes.

8.6 RESULTS

In this section, numerical results are presented to demonstrate the validity of the proposed filter and the performance monitoring schemes. The data for the demonstration was generated using a battery simulator represented in Figure 6.3. The battery simulator uses the equivalent circuit model shown in Figure 8.5(b) to simulate the voltage and current measurements that resample real-time measurements from a battery. The OCV effect of the battery, denoted by $V_o(s(k))$ in Figure 8.5(b), was generated using the combined+3 model with the following model parameters: $k_0 = -9.082$, $k_1 = 103.087$, $k_2 = -18.185$, $k_3 = 2.062$, $k_4 = -0.102$, $k_5 = -76.604$, $k_6 = 141.199$, and $k_7 = -1.117$. The voltage measurements across the battery were simulated using this observation model (8.21) and the current measurement model was implemented after (8.20). The relaxation parameters of the ECM are set at $R_0 = 0.2$, $R_1 = 0.1$, $C_1 = 2$, $R_2 = 0.3$, and $C_2 = 5$. The electrical ECM model in the battery simulator can be changed in a way that the RC models can be selected, from the set of $\{(R_1, C_1), (R_2, C_2)\}$.

Figure 8.6 shows the simulated voltage and current, $z_v(k)$ and $z_i(k)$, respectively, obtained from the battery simulator for current and voltage measurement noise $\sigma_v = \sigma_i = 0.01$.

The EKF approach for SOC estimation is implemented by assuming (8.19) as the process model and (8.21) as the measurement model. This model assumes that there is no uncertainty in the knowledge of the battery capacity and the internal resistance. The process noise for this model is $\Delta_k^2 \sigma_i^2 / 3600 \hat{Q}^2$ and the measurement noise is $\sigma_v^2 + \hat{R}_0^2 \sigma_i^2$.

First, the standard EKF algorithm was demonstrated without introducing any model uncertainties (i.e., the EKF algorithms used the same values of Q and R_0 used by the battery simulator). Figure 8.7 shows the estimated SOC by the EKF algorithm for the noise scenario shown in Figure 8.6. For comparison, a Coulomb counting-based SOC estimation method is implemented based on (8.17) where the measured current $z(k)$ is used instead of $i(k)$. The true SOC is computed by the battery simulator based on (8.17) with the noiseless current values $i(k)$ and plotted in the same figure for comparison. It can be observed from Figure 8.7 that the variance of the SOC estimates by the EKF increases with the noise standard deviations.

Figure 8.8 shows one of the many advantages of the EKF approaches to SOC estimation. Here, the EKF approach and the conventional Coulomb counting approaches

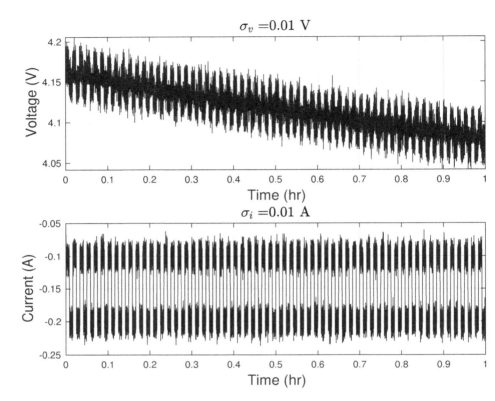

Figure 8.6 Voltage and current simulated with measurement noise.

were given the worst possible initial SOC information on SOC (i.e., both the CC approach and the EKF approach assume that the initial SOC is 0.5). The simulation demonstrates the advantages of the EKF approach which quickly recovers to the correct SOC. However, the Coulomb counting carries the initial error throughout; this demonstrates the advantage of the EKF in practical applications.

Figure 8.9(a) shows the performance of the EKF when the parameters of the state-space model are perfectly known and the s.d. of two noises are $\sigma_v = 0.01\text{V}$ and $\sigma_i = 0.01\text{A}$. Figure 8.9(b) shows the average NIS values and the corresponding confidence limits computed based on the approach described in Section 8.5.1. The average NIS compliance is computed as the percentage of the average NIS values that

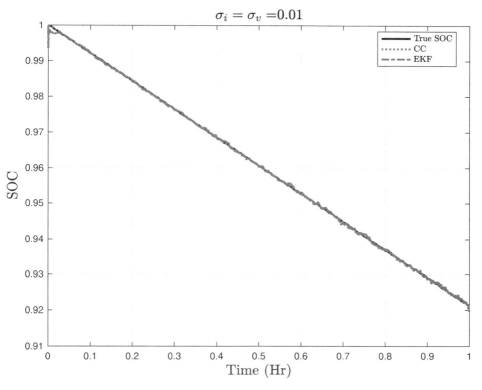

Figure 8.7 SOC tracking for at $\sigma_v = \sigma_i = 0.01$ noise variances.

are within the bounds. Similarly, Figure 8.9(c) shows the averaged t-statistics and the corresponding limits computed based on Section 8.5.2.

Figure 8.10(a) shows the performance of the standard EKF when model uncertainty is introduced in the process noise variance. Here, voltage and current measurement noises had s.d. of $\sigma_v = 0.01$V and $\sigma_i = 0.01$A, respectively. Here, the EKF assumed an incorrect voltage measurement noise that is 10 times the true standard deviation of the noise. The averaged NIS and t-statistics are within the confidence bounds indicating that the t-statistic is unable to detect model uncertainty in the measurement variance. However, uncertainties in the measurement and process noise variances are the most common and acceptable form of filter mismatches. These mismatches result in added variance in the estimates.

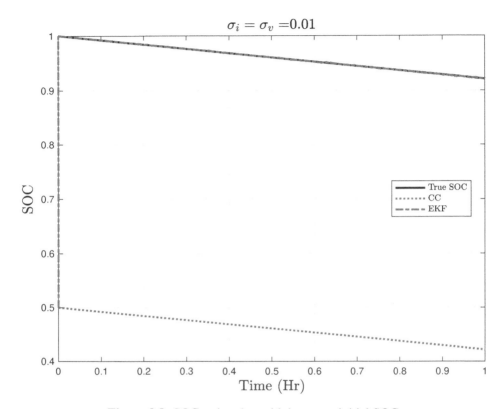

$$\sigma_i = \sigma_v = 0.01$$

Figure 8.8 SOC estimation with incorrect initial SOC.

Figure 8.11(a) shows the results of mismatched EKF when the ECM parameter R_0 assumed in the filter is different from the one used in the battery simulator. Here, the EKF assumed an incorrect R_0 that is twice the value of the true R_0. Except for the mismatch in R_0, all the parameters available to the EKF are the same as the ones used in the battery simulator. The averaged NIS and t-statistics consistently indicate the mismatch in the filter, as shown in Figure 8.11(b, c).

Figure 8.12(a) shows the results of mismatched EKF when the battery capacity assumed in the filter is different from the one used in the battery simulator. Here, the EKF assumed an incorrect $\hat{Q}$ that had $\Delta_Q = -0.5$ Ah. The averaged NIS and t-statistics consistently indicate the mismatch in the filter, as shown in Figure 8.12(b, c).

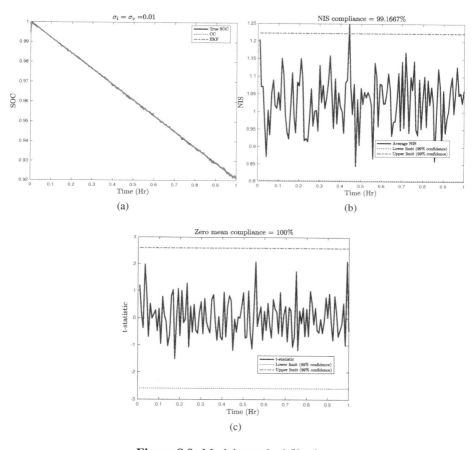

Figure 8.9 Model-matched filtering.

8.7 CONCLUSIONS

The battery fuel gauge is the most crucial part of a battery management system and the SOC is the most important quantity to be estimated by a BFG. This chapter briefly summarized the three well-known approaches to SOC estimation: the Coulomb counting approach, the voltage lookup-based approach, and the fusion-based approach, and discussed their limitations. The fusion-based approach to SOC estimation, based on

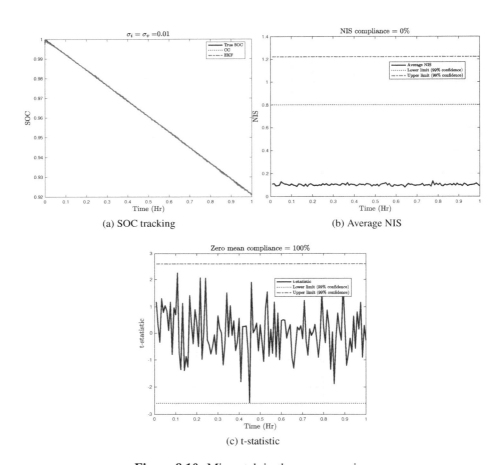

(a) SOC tracking   (b) Average NIS

(c) t-statistic

Figure 8.10 Mismatch in the process noise.

the EKF, was discussed in detail. Even though the EKF approach to SOC estimation has received wider attention in the literature, its adaptation is sparse due to the fact that EKF is reliable only when the model parameters are accurately known. When there is a discrepancy between the underlying model parameters and its assumption, the EKF starts to diverge. This chapter provided insights into filter divergence and ways to monitor it.

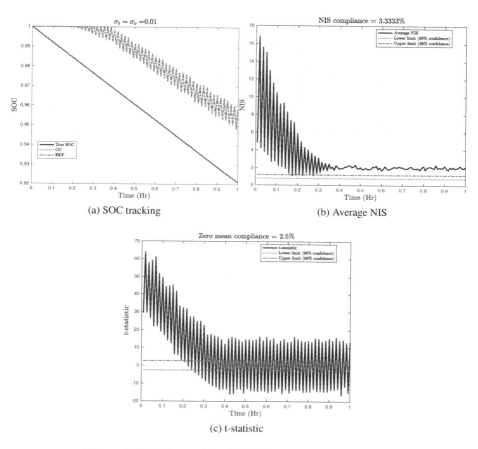

(a) SOC tracking

(b) Average NIS

(c) t-statistic

Figure 8.11 Mismatch in the ECM model parameter R_0.

It must be reemphasized that the EKF approach presented in this chapter is only a demonstration based on simulated data. The state-space model in this chapter assumed that the model parameters were time-invariant and were known. In reality, the model parameters need to be estimated and their variability with time (or SOC) must be carefully accounted for. Chapter 2 provided several examples to understand the effect of model mismatch and ways to observe them. The model mismatch analysis presented in Chapter 2 provided a guideline for designing the fusion-based SOC estimation

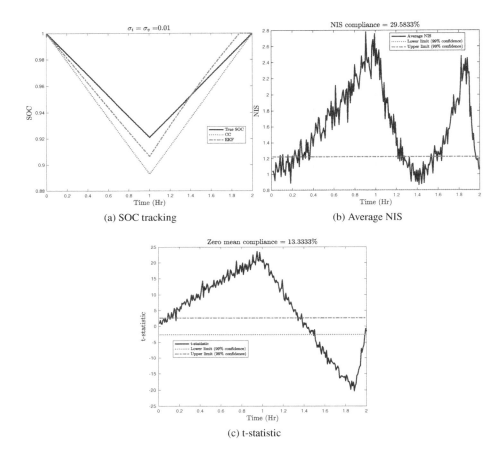

(a) SOC tracking   (b) Average NIS

(c) t-statistic

Figure 8.12 Mismatch in the battery capacity Q.

algorithms where the state-space model consists of close to a dozen model parameters, such as the internal resistance, RC elements, battery capacity, hysteresis parameters, and measurement noise statistics. The EKF-based approach to SOC estimation has been the subject of intense research and several approaches were presented to tackle the model uncertainty problem. To date, no approach has emerged as a universally accepted EKF framework for SOC estimation. Readers are encouraged to do their own research about the various EKF-based SOC estimation approaches found in the literature. When evaluating the existing approaches, the model mismatch analysis presented in Chapter 2 can be utilized as a tool for evaluating them. The reader is also encouraged to read more about filter consistency testing in [5].

8.8 BIBLIOGRAPHICAL NOTES

The nonlinear filtering approach to SOC estimation using the extended Kalman filter was first reported by G.L. Plett [1]. The improved process model presented in this chapter for SOC estimation was based on the work presented in [4]. More technical details of the extended Kalman filter can be found in [5].

References

[1] G.L. Plett, "Extended Kalman filtering for battery management systems of LiPB-based HEV battery packs: Part 1. Background," *Journal of Power Sources,* Vol. 134, No. 2, pp. 252–261, 2004.

[2] B. Balasingam, and P. Kumar, "Battery fuel gauge: A crucial element in Battery Management Systems," *IEEE Instrumentation & Measurement Magazine,* Vol. 25, No. 7, pp. 14–20, 2022.

[3] M.S. Ahmed, S.A. Raihan, and B. Balasingam, "A scaling approach for improved state of charge representation in rechargeable batteries," *Applied Energy,* Vol. 267, pp. 114880, 2020.

[4] K. Movassagh, A. Raihan, B. Balasingam, and K. Pattipati, "A critical look at Coulomb Counting approach for state of charge estimation in batteries," *Energies,* Vol. 14, No. 14, pp. 4074, 2021.

[5] Y. Bar-Shalom, X.R. Li, and T. Kirubarajan, *Estimation with applications to Tracking and Navigation: Theory, Algorithms and Software,* John Wiley & Sons, New York, 2004.

[6] P. Pillai, S. Sundaresan, K.R. Pattipati, and B. Balasingam, "Optimizing current profiles for efficient online estimation of battery equivalent circuit model parameters based on Cramer-Rao lower bound," *Energies,* Vol. 15, No. 22, pp. 8441, 2022.

Chapter 9

Battery Thermal Management

9.1 INTRODUCTION

Temperature management is crucial to the performance of Li-ion batteries. At lower temperatures, the battery impedance increases and results in reduced output power. At higher temperatures, the battery is susceptible to triggering thermal runaway. For safe and reliable performance, the working temperature of a Li-ion battery should be in the range of 15–35°C. A battery thermal management system (BTMS) is tasked with ensuring that the battery pack stays within allowable temperature limits. A typical BTMS consists of temperature sensors across the surface of a battery pack and cooling and heating systems that are activated based on the measured and predicted temperature of the battery. Most BTMS use mediums such as air, liquid, and phase change materials (PSM) to transport heat from/to battery cells. The next section briefly reviews three different mediums used in battery thermal management.

9.2 THERMAL MANAGEMENT MEDIUMS

9.2.1 Air

Air cooling is the simplest form of cooling adopted in several electric vehicle models, such as the Toyota Prius (2016–present), Nissan Leaf (2018–present), and BYD E6 (2009–present). Air cooling can be further divided into forced and natural air cooling; in forced air cooling (see Figure 9.1(c)), fans are used to induce the airflow, whereas in natural air cooling no fans are used. Air cooling has limitations due to the low thermal conductivity of air, which can be improved by modification in channel design and cell arrangement.

Different airflow channels were also experimented with to improve cooling efficiency. For example, in comparison to a simple channel, the cooling is effective in a wedge channel (see Figure 9.1(b)) where the area in the wedge channel is gradually decreased from inlet to outlet, which gives the batteries at the outlet a good chance to receive enough air for cooling. Figure 9.1 shows different types of cooling channel arrangements to analyze the heat transfer performance.

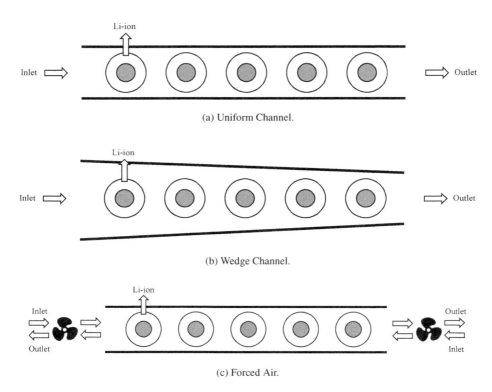

Figure 9.1   (a–c) Cooling channel arrangement.

It was found that temperature uniformity can also be improved by the arrangement of cells in a battery pack. Figure 9.2 shows three arrangements of Li-ion cells in the battery pack. Aligned and staggered arrangements were found to be more prone to high-temperature rise and nonuniformity in the Li-ion cells, whereas the trapezoidal arrangement experienced better temperature uniformity.

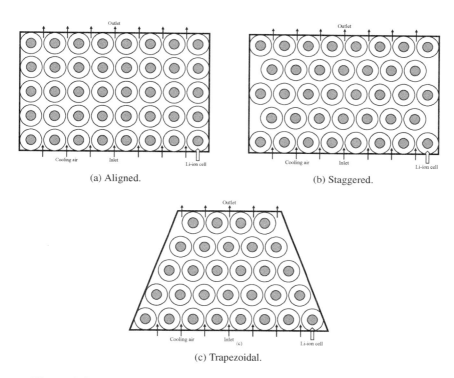

(a) Aligned.

(b) Staggered.

(c) Trapezoidal.

Figure 9.2 (a–c) Air cooling with different cylindrical cells arrangement.

There are several positive attributes of air as a thermal management medium: it does not require much space, it is low cost, and the design is very simple. Many practical systems, particularly electronic circuit boards, use air as a thermal management medium. However, air as a thermal management medium suffers from low heat capacity when it comes to managing large battery packs. Liquids, on the other hand, possess much higher heat capacity than air.

9.2.2 Liquid

Liquids have better thermal conductivity than air. Liquid cooling can be divided into direct and indirect cooling. In direct cooling, batteries are kept in direct contact with the liquid coolant. In indirect cooling, coolant flows in tubes, cold plates, and jackets that are in contact with the surface of the batteries. The advantage of direct cooling is that the heat transfer rate is high due to the direct contact battery surface with the coolant. The disadvantages of direct cooling are the safety risks and state of health deterioration. In practical thermal management applications, a direct cooling strategy is used in high-performance vehicles, such as sports cars. In passenger electric vehicles, where safety and state of health requirements are high, indirect cooling strategies are employed.

Figure 9.3 shows how cooling plates, embedded with cooling channels, are used in battery thermal management. The cooling plate helps to extract heat from the battery and to transfer it to the coolant. Cooling plates should have good thermal conductivity, low weight, and high flexibility. The contact of the channels with the cooling plate can be increased by channel design. Figure 9.4 shows three different types of channel design.

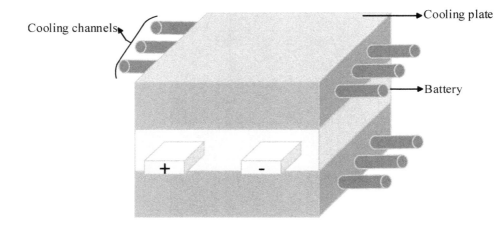

Figure 9.3 Cooling plate.

Compared to air, liquid coolants have higher specific heat and thermal conductivity that results in better cooling efficiency. However, liquid cooling systems require a more careful and complex design. Improving the efficiency of a battery cooling system remains an ongoing research topic. One aspect of this research involves improving the thermo-physical properties of coolant fluids. In order to be effective, the coolants need to have a high thermal capacity, less viscosity, low density, and low freezing temperature. Nanomaterials are being investigated to improve the required properties of coolants.

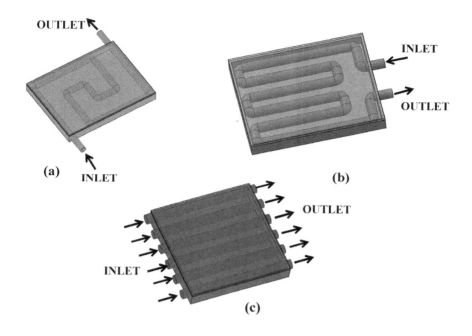

Figure 9.4 Cooling channel designs: (a) serpentine, (b) U-turn, and (c) multichannel.

9.2.3 Phase Change Material

Phase change materials (PCMs) have the ability to regulate heat. When exposed to excessive heat, the PCM undergoes a phase change and absorbs the heat energy as a result. When the temperature drops, the phase change reverses and the PCM releases heat. Heat regulation using PCM does not require any external inputs; the heat regulation happens naturally due to the property of the PCM. As such, PCM-based cooling is considered to be a passive cooling strategy. PCMs have been used for thermal management in wide-ranging applications, such as buildings, electronic parts, and EVs. In order to be used in electric vehicles, PCMs need to have certain qualities: good thermal conductivity, low volume expansion, phase stability, and low flammability are a few of them. Paraffin wax, Pentacosane, Polyethylene glycol (PEG) 1000, Phase Change

Component (PCC) 37, PCC48, and PCC55 are some of the examples of PCM that can be used in EVs.

The challenge with PCMs is their low thermal conductivity. It was shown that the thermal conductivity of PCMs can be improved by mixing nanoparticles with pure PCM. Also, the performance of a pure PCM-based system can be further improved by using it with some active cooling techniques. Figure 9.5 shows the concept of a mixed cooling system consisting of PCMs and forced air cooling for cylindrical batteries. In this setup, the batteries are immersed within the PCM, which functions both as a heat regulator (inherent nature of PCMs) and as a heat conductor. The excess heat from the battery is transferred by the PCM to the fins and then transferred away by the forced air flowing through the fins.

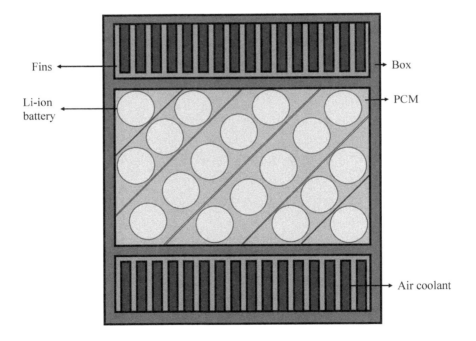

Figure 9.5 Combined PCM and air cooling.

9.3 BATTERY THERMAL MODELING

Battery thermal models are useful to estimate the amount of temperature increase in batteries for given charging and discharging scenarios. The battery thermal management system needs to be designed based on the predicted/expected temperature rise scenario during the lifetime of a battery. This section describes battery thermal models and an approach to predict the temperature rise based on this model.

The thermal-electric equivalent circuit model (TECM) approximates the thermal behavior of the battery. Table 9.1 compares the electrical properties and their thermal equivalent.

Table 9.1
Thermal Equivalence of Electrical Properties

Electrical Property	Unit	Thermal Equivalent Property	Unit
Electron flow	A	Heat flow	W
Electron potential	V	Temperature	K
Electric resistance	Ω	Thermal resistance	K/W
Electric capacitance	F	Thermal capacitance	J/W

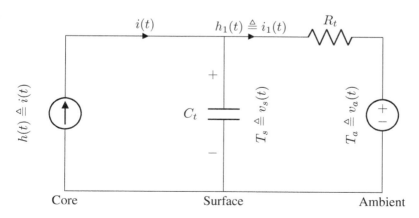

Figure 9.6 TECM.

Figure 9.6 shows a simplified one-point version of a TECM. The heat flow $h(t) \equiv i(t)$ is analogous to electric current; measured surface temperature $v_s(t)$, and ambient temperature $v_a(t)$ are analogous to voltage; thermal resistance R_t is analogous to electric resistance; and thermal capacitance C_t is analogous to electrical capacitance. Also,

- The generated heat $h(t)$ acts analogous to a current source in the electrical domain. The generated heat can be approximated as [1]

$$h(t) = \underbrace{i_e{}^2(t)R_e(t)}_{\text{irreversible heat}} + \underbrace{i_e T \frac{dV_o}{dT}}_{\text{reversible heat}} \tag{9.1}$$

where $i_e(t)$ is the electrical current through the battery at time t, $R_e(t)$ is the electrical internal resistance of the battery (assuming the R-int electrical equivalent circuit model) at time t, V_o is the open circuit voltage of the battery, T is the temperature, and $\frac{dV_o}{dT}$ is also known as the entropic heat coefficient.

- The ambient temperature $v_a(t)$ acts analogous to a voltage source (i.e., voltage sink) in the electrical domain.

Here, the battery is considered as a single point, that is, the temperatures (surface temperature $v_s(t)$ and the ambient temperature $v_a(t)$) denote the average temperature along the surface of the battery. In a real battery, the temperature may not be equally distributed along its surface. The advanced model must take this factor into account when developing battery thermal management systems.

The current through capacitance C_t can be written as

$$C_t \dot{T}_s(t) = h(t) - h_1(t) = h(t) - \frac{T_s(t) - T_a(t)}{R_t} \tag{9.2}$$

$$\dot{T}_s(t) = -\left(\frac{1}{R_t C_t}\right) T_s(t) + \left(\frac{1}{R_t C_t}\right) T_a(t) + \frac{h(t)}{C_t} \tag{9.3}$$

By discretizing (9.3) (see Appendix C for details), we get

$$T_s[k+1] = \alpha T_s[k] + (1-\alpha)T_a[k] + (1-\alpha)R_t\, h[k] \tag{9.4}$$

where k denotes the time index, $\alpha = e^{-\Delta/R_t C_t}$, and Δ is the sampling time. By substituting $k \leftarrow k-1$ in (9.4), we get

$$T_s[k] = \alpha T_s[k-1] + (1-\alpha)T_a[k-1] + (1-\alpha)R_t\, h[k-1] \tag{9.5}$$

It can be noticed that (9.5) can be written in vector form as

$$T_s[k] = \mathbf{a}[k]^T \mathbf{b} \tag{9.6}$$

where the observation model $\mathbf{a}[k]^T$ and the model parameter vector $\mathbf{b}$ are given by

$$\mathbf{a}[k]^T \triangleq \left[T_s[k-1], \; T_a[k-1], \; h[k-1] \right] \tag{9.7}$$

$$\mathbf{b} \triangleq \left[b_1, \; b_2, \; b_3 \right]^T \tag{9.8}$$

and the elements of $\mathbf{b}$ are given by $b_1 = \alpha$, $b_2 = 1 - \alpha$, and $b_3 = (1 - \alpha)R_t$.

Now the parameter estimation approach is modified to formally account for the measurement noise. Let us model the measured ambient noise as

$$z_a[k] = T_a[k] + n_a[k] \tag{9.9}$$

where $z_a[k]$ denotes the measured ambient temperature and $n_a[k]$ denotes the measurement noise that is assumed zero-mean i.i.d. Gaussian with standard deviation σ_a. Similarly, the measured surface temperature and heat flow are written, respectively, as

$$z_s[k] = T_s[k] + n_s[k] \tag{9.10}$$

$$z_h[k] = h[k] + n_h[k] \tag{9.11}$$

where $n_s[k]$ and $n_h[k]$ are the corresponding measurement noises that are assumed zero-mean white with standard deviations σ_s and σ_h, respectively.

By substituting $z_s[k]$, $z_a[k]$, and $z_h[k]$ for $T_s[k]$, $T_a[k]$, and $h[k]$, respectively, in (9.5), we can rewrite it as:

$$\begin{aligned} z_s[k] &= \alpha z_s[k-1] + (1-\alpha)z_a[k-1] + (1-\alpha)R_t z_h[k-1] + n_s[k] \\ &\quad - \alpha n_s[k-1] - (1-\alpha)n_a[k-1] - (1-\alpha)R_t n_h[k-1] \\ &= \mathbf{a}[k]^T \mathbf{b} + n[k] \end{aligned} \tag{9.12}$$

where the observation model $\mathbf{a}[k]^T$ is analogous to $\mathbf{a}[k]$ in (9.7)

$$\mathbf{a}[k]^T \triangleq \left[z_s[k-1], \; z_a[k-1], \; z_h[k-1] \right] \tag{9.13}$$

The noise in the surface temperature observation (9.12) can be given by

$$n[k] \triangleq \overline{n}_s[k] + \overline{n}_a[k] + \overline{n}_h[k] \tag{9.14}$$

where $\bar{n}_s[k] = n_s[k] - \alpha n_s[k-1]$, $\bar{n}_a[k] = -(1-\alpha)n_a[k-1]$, and $\bar{n}_i[k] = -(1-\alpha)R_t n_h[k-1]$.

The autocorrelation of the noise $n[k]$ is written as

$$\mathcal{R}_n(l) = E\{n[k]n[k-l]\} \tag{9.15}$$

and it can be shown that

$$\mathcal{R}_n(l) = \begin{cases} \mathcal{R}_n(0) & l = 0 \\ \mathcal{R}_n(1) & l = 1 \\ 0 & l > 1 \end{cases} \tag{9.16}$$

where

$$\mathcal{R}_n(0) = (1+\alpha)^2\sigma_s{}^2 + (1-\alpha)^2\sigma_a{}^2 + (1-\alpha)^2 R_t^2\sigma_h{}^2 \tag{9.17}$$

$$\mathcal{R}_n(1) = -\alpha\sigma_s{}^2 \tag{9.18}$$

By considering a batch of L_b observations, (9.12) can be rewritten in matrix form as

$$\mathbf{z}_s = \mathbf{Ab} + \mathbf{n} \tag{9.19}$$

where

$$\mathbf{z}_s = [z_s(1), z_s(2), \dots, z_s(L_b)]^T, \tag{9.20}$$
$$\mathbf{A} = [\mathbf{a}(1), \mathbf{a}(2), \dots, \mathbf{a}(L_b)]^T$$

and the $L_b \times L_b$ covariance of the noise vector $\mathbf{n}$ can be defined as

$$\mathbf{R} = E\{\mathbf{nn}^T\} \tag{9.21}$$

where $\mathbf{R}$ is a tridiagonal Toeplitz matrix with diagonal and off-diagonal elements given by $\mathcal{R}(0)$ and $\mathcal{R}(1)$, respectively (see (9.17) and (9.18)).

Considering the model as shown in (9.19), the parameter $\mathbf{b}$ can be estimated using the least squares method as follows:

$$\hat{\mathbf{b}} = (\mathbf{A}^T\mathbf{R}^{-1}\mathbf{A})^{-1}\mathbf{A}^T\mathbf{R}^{-1}\mathbf{z}_s \tag{9.22}$$

Now the thermal model parameters can be recovered as

$$\hat{R}_t = \frac{\hat{b}_3}{(1-\hat{b}_1)}, \quad \hat{C}_t = \frac{-\Delta}{\log(\hat{b}_1)\hat{R}_t} \tag{9.23}$$

where $\hat{\mathbf{b}} = \left[\hat{b}_1, \hat{b}_2, \hat{b}_3\right]^T$.

9.4   SIMULATION RESULTS

In this section, the thermal modeling and parameter estimation approach developed in Section 9.3 is evaluated through simulation analysis.

The thermal model parameters are the heat resistance R_t and the heat capacitance C_t shown in Figure 9.6. The goal of the experiment in this section is to demonstrate the estimation of these two parameters and battery surface temperature prediction based on three measurements: heat flow $z_h[k]$, surface temperature $z_s[k]$, and ambient temperature $z_a[k]$. The performance of the algorithms will be evaluated at different noise levels. In order to do that, the signal-to-noise ratio of each measurement is defined as

$$\text{SNR} = 10 \log \left(\frac{P_{\text{signal}}}{P_{\text{noise}}} \right) \tag{9.24}$$

where P_{signal} denotes the power of the signal and P_{noise} denotes noise power. The SNR of each measurement $z_h[k]$, $z_s[k]$, and $z_a[k]$, respectively, are

$$\text{SNR}_h = 10 \log \left(\frac{P_h}{\sigma_h^2} \right), \quad \text{SNR}_s = 10 \log \left(\frac{P_s}{\sigma_s^2} \right), \quad \text{SNR}_a = 10 \log \left(\frac{P_a}{\sigma_a^2} \right)$$

where P_h, P_s, and P_a denote the average power that is defined for a generic signal $x(t)$ as follows:

$$P_x = \frac{1}{T} \int_0^T x^2(t) dt \tag{9.25}$$

It is assumed that the measurement uncertainty is the same in all measured quantities; this would result in

$$\text{SNR}_h = \text{SNR}_s = \text{SNR}_a \tag{9.26}$$

With the above assumptions, the measurement simulation was done considering a battery with a capacity of 3000 mAh and true internal resistance of 0.04Ω. First, a 2000-second current profile, as shown in the top plot of Figure 9.7, is generated. Assuming a simple R-int model (see Chapter 3) to represent the electrical behavior of the battery with parameter $R_e = 0.04\,\Omega$, the heat generated in the battery, $h(t)$, is simulated and Figure 9.7 (bottom) shows one such realization of the generated heat.

Assuming $R_t = 10$ K/W and $C_t = 100$ J/K for the true values of the thermal resistance and capacitance, respectively, and assuming the ambient temperature to be

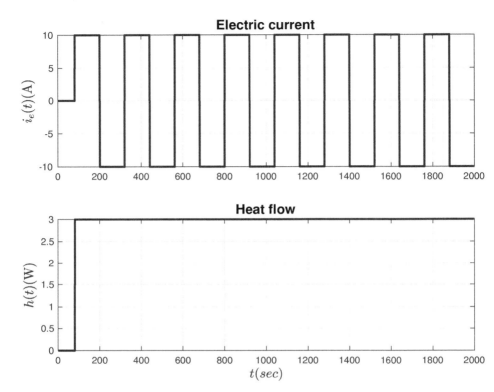

Figure 9.7 Heat generation in the battery. The top plot shows the current through the battery and the plot at the bottom shows the generated heat in watts.

$v_a = 298.15\text{K} \,(25^\circ\text{C})$ the true surface temperature of the battery $v_s[k]$ is computed using (9.5). Then, the measurement models (9.9), (9.10), and (9.11) are used to simulate measured data, $z_a[k]$, $z_s[k]$, and $z_h[k]$ for a certain SNR. Figure 9.8 shows simulated heat flux, $z_h(t)$, surface temperature, $z_s(t)$, and ambient temperature $z_a(t)$ for two different SNR values.

The simulated measurements, $z_a[k], z_s[k]$, and $z_h[k]$, shown in Figure 9.8, were used to estimate the TECM parameters of the battery. The estimated parameters are then used to predict the temperature during a battery discharge test.

The battery discharge data is simulated at a 2C-discharge rate as shown in Figure 9.9. This simulation was done considering true values of thermal parameters, and

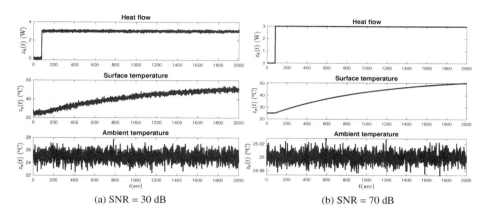

(a) SNR = 30 dB   (b) SNR = 70 dB

Figure 9.8 (a, b) Simulated measurements for two different SNR values.

internal resistance with no noise. The simulated surface temperature shown in Figure 9.9 will be used later for comparison with the predicted temperature based on the estimated parameters.

Four different values of SNR 40, 50, 60, and 70 dB are chosen to simulate the battery measurements similar to the one shown in Figure 9.8. Figure 9.10 shows the temperature prediction when the TECM model parameter is perfectly known.

In a real-world scenario, the internal resistance of the battery is not known. Figure 9.11 shows the temperature prediction when there is a specific uncertainty in the electric ECM model. Particularly, the prediction algorithm assumed that the internal resistance $R_e = 0.03\Omega$, whereas the true value used to simulate the surface temperature is $R_e = 0.04\Omega$.

9.5 CONCLUSIONS

This chapter provided a brief review of battery thermal management strategies and an approach to predict the temperature rise in battery packs by estimating its thermal-electrical equivalent-circuit model parameters. The modeling approach consists of a training period in which the voltage, current, and temperature data are collected to estimate the model parameters: electrical resistance, thermal resistance, and thermal capacitance. This approach can predict the temperature rise on the battery surface for given and expected current profiles.

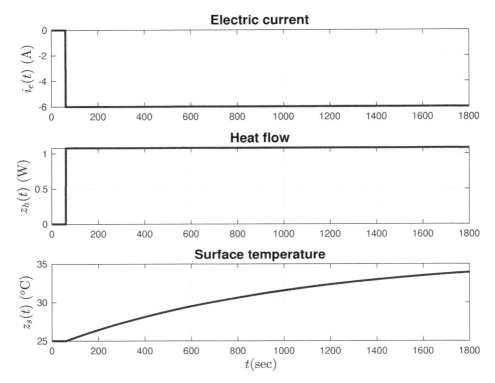

Figure 9.9 Discharge measurements. Simulated battery discharge measurements at a 2C rate considering true values of thermal parameters, and internal resistance. No noise was included in the simulation.

9.6 BIBLIOGRAPHICAL NOTES

Battery thermal management is a new and growing research topic. Some discussion shown in Section 9.2 referred to [2–4] and the references therein. The reader is referred to [1] for more details on the application of the temperature prediction approach to real battery data.

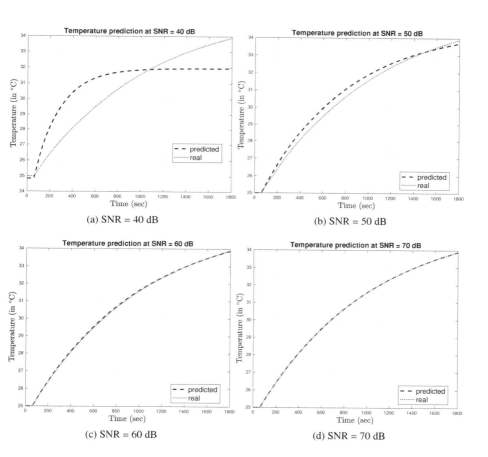

(a) SNR = 40 dB

(b) SNR = 50 dB

(c) SNR = 60 dB

(d) SNR = 70 dB

Figure 9.10 (a–d) Surface temperature prediction using ideal electric ECM.

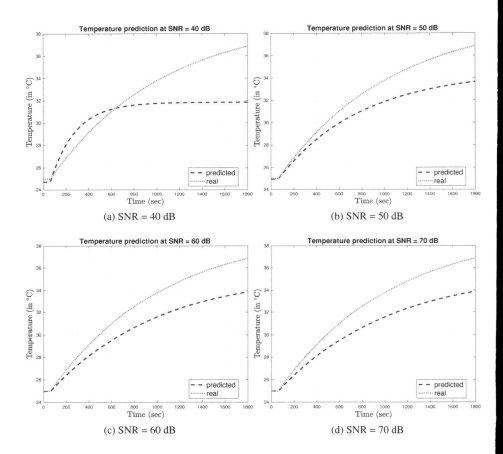

Figure 9.11 (a–d) Surface temperature prediction using the electric ECM with uncertainty.

References

[1] P. Kumar, G. Rankin, K.R. Pattipati, and B. Balasingam, "Model based approach to long term prediction of battery surface temperature," *IEEE Journal of Emerging and Selected Topics in Industrial Electronics,* 2022.

[2] Z. Rao, and S. Wang, "A review of power battery thermal energy management," *Renewable and Sustainable Energy Reviews,* Vol. 15, No. 9, pp. 4554–4571, 2011.

[3] G. Xia, L. Cao, and G. Bi, "A review on battery thermal management in electric vehicle application," *Journal of Power Sources,* Vol. 367, pp. 90–105, 2017.

[4] J. Kim, J. Oh, and H. Lee, "Review on battery thermal management system for electric vehicles," *Applied Thermal Engineering,* Vol. 149, pp. 192–212, 2019.

Chapter 10

Optimal Charging Algorithms

10.1 INTRODUCTION

This chapter explains the fundamentals of Li-ion battery charging during which the depleted electrons in the battery are refilled. Charging is accomplished by connecting the battery either to a constant voltage (CV) power supply or a constant current (CC) power supply. However, when it comes to Li-ion battery charging, greater attention must be paid to the charging current and voltage because the batteries are very sensitive to over-voltage and excess temperature during charging. Fast charging is a preference in many applications, such as EVs; however, a high charging current causes energy loss and excessive heat. Hence, the charging current should be selected in a way that both the charging time and energy loss (or excessive heat) are minimized. This chapter details how these two competing objectives can be managed through optimization.

10.2 CHARGING STRATEGIES

In this section, several charging strategies are reviewed and discussed.

10.2.1 Constant Current Charging

CC charging maintains a specific charging current i_{CC}. Figure 10.1 depicts a CC charging scenario for a battery represented by R-int equivalent circuit model. The charging process is terminated (or shut down) when the terminal voltage reaches $v = \mathrm{OCV_{max}}$.

$$\mathrm{OCV_{max}} = V_o(s_{\mathrm{sd}}) + i_{\mathrm{CC}}R_0 \qquad (10.1)$$

where $s \triangleq s_{\mathrm{sd}}$ is the SOC when the battery shuts down as the terminal voltage reaches $\mathrm{OCV}_{\max}$.

The shutdown SOC can be computed as

$$s_{\mathrm{sd}} = f^{-1}(\mathrm{OCV}_{\max} - i_{\mathrm{CC}} R_0) \tag{10.2}$$

where $f^{-1}(\cdot)$ denotes the inverse of the OCV function. Under the CC topology, the battery SOC increases at a fixed rate of i_{CC}/Q where Q is the battery capacity.

It is easy to see from (10.2) that the higher the charging current the lower the final SOC. In order to fully charge the battery (i.e., to bring the SOC to 1), the current must approach zero; this is accomplished by the CV charging during which the current gradually reduces.

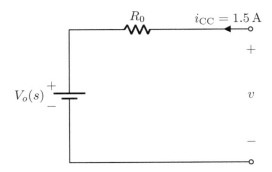

Figure 10.1 Constant current charging scenario.

Example 10.1

Consider the constant current charging scenario shown in Figure 10.1. The OCV of the battery is defined by the combined+3 model

$$V_0(s) = k_0 + k_1 s^{-1} + k_2 s^{-2} + k_3 s^{-3} + k_4 s^{-4}$$
$$+ k_5 s + k_6 \ln(s) + k_7 \ln(1-s) \tag{10.3}$$

where $k_0 = -9.081846$, $k_1 = 103.087009$, $k_2 = -18.184590$, $k_3 = 2.062476$, $k_4 = -0.101779$, $k_5 = -76.603691$, $k_6 = 141.199419$, and $k_7 = -1.116841$. These parameters were computed for SOC values $s \in [0, 1]$ that were linearly scaled to $s' \in [0.175, 0.825]$. The internal resistance of the battery is $R_0 = 0.2\Omega$ and the battery capacity is $Q = 1.5$ Ah.

- Assuming that the battery was empty at the start, compute the charging time using constant current $i_{CC} = 1.5$A.

- What is the SOC at the end of charging?

- Compute and plot the terminal voltage v across the battery and SOC throughout the charging period.

The maximum OCV can be computed by substituting $s = 1 - \epsilon = 0.825$ in (10.3) to be

$$\text{OCV}_{\text{max}} = V_0(0.825) = 4.1937 \text{ V} \tag{10.4}$$

The voltage drop is

$$v_{\text{d}} = i * R_0 = 1.5 * 0.2 = 0.3 \text{ V} \tag{10.5}$$

Hence, the OCV at the end of charging (or shut down) is

$$\text{OCV}_{\text{sd}} = \text{OCV}_{\text{max}} - v_{\text{d}} = 3.8937 \text{ V} \tag{10.6}$$

The SOC and the end of charging can be computed by

$$\text{SOC}_{\text{sd}} = f^{-1}(\text{OCV}_{\text{sd}}) = f^{-1}(3.8937) = 0.6361 \tag{10.7}$$

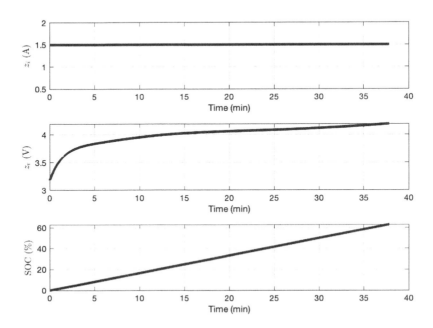

Figure 10.2 Constant current charging. The SOC increases at a constant rate during CC charging. The higher the current the lower the SOC at the end of charging.

One can notice that the battery is not fully charged. Assuming that the initial SOC is zero, the charging time is

$$t_c = \frac{\text{SOC}_{\text{sd}} * Q}{i_{\text{cc}}} = \frac{0.6361 * 1.5}{1.5} = 0.6361\,\text{hr} \approx 38\,\text{min}. \qquad (10.8)$$

Figure 10.2 shows the voltage and SOC of the CC entire charging described in Example 10.1.

10.2.2 Constant Voltage Charging

It was explained through Example 10.1 in Section 10.2.1 that the CC charging strategy will not be able to fully charge the battery. In order to fully charge the battery, the charging mode must be changed to CV. Figure 10.3 shows a circuit diagram during CV charging. The charger is kept at a constant voltage of $v_{CV} = OCV_{max}$, the maximum allowable limit. Here, the charging current will gradually decrease. The charging process is terminated when the charging current falls below a certain threshold.

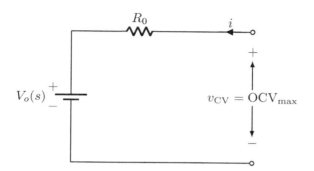

Figure 10.3 Constant voltage charging scenario.

The CV charging strategy is suitable to fully charge the battery and to bring its SOC to 1. However, the CV strategy is not suitable at the initial stages of charging where the charging current can be excessively high. Consider the CV charging scenario shown in Figure 10.3. The CV charging current is given by $i_{cv} = (OCV_{max} - f(s_0))/R_0$ where s_0 denotes the initial SOC. The charging current during the CV approach entirely depends on s_0 and R_0, whereas many Li-ion batteries require limits on the charging current to ensure the safety and reliability of the battery.

Figure 10.4 shows the current, voltage, and SOC during constant voltage charging. The same parameters used in Example 10.1 were used to simulate Figure 10.4.

10.2.3 Constant Current-Constant Voltage Charging

The CC charging strategy (Section 10.2.1) is unable to fully charge the battery. The CV charging (Section 10.2.2) is able to fully charge the battery; however, it may lead to an initial charging current above allowable limits. The CC-CV charging strategy is a combination of both CC and CV strategies. Initially, the battery is charged with CC strategy allowing it to control the initial current within allowable limits. After that, the charging mode is switched to CV to fully charge the battery.

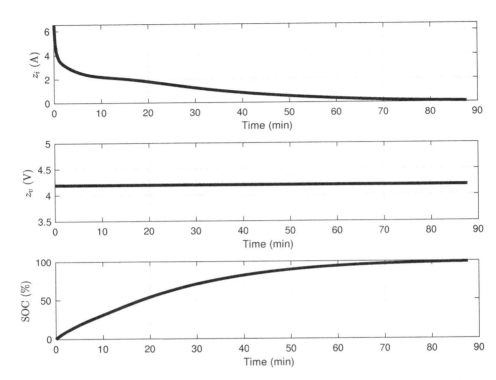

Figure 10.4 Constant voltage charging. Sample current and voltage profile of CV charging. Parameters from Example 10.1 were used to simulate the CV charging of a battery from empty. The charger is terminated when the SOC reaches 99%. It took 90 minutes to charge the battery.

Remark 10.1 The CC strategy allows one to select the initial charging current that is safe enough not to cause severe lithium plating and temperature increase. By selecting the initial current at a desired level, one can control the SOH degradation of the battery.

Example 10.2

Consider the battery parameters introduced in Example 10.1. This battery is charged according to the following protocol.

- Constant current charging with 1.5A until the voltage reaches OCV_{max}.

- When the terminal voltage is $v = OCV_{max}$, switch to constant voltage charging.

- Terminate charging when SOC reaches 99%.

Compute the voltage, current, and SOC throughout the charging duration and plot them. Assume that the battery is empty at the start.

Figure 10.5 shows the voltage, current, and SOC curves. The total charging time was 101 minutes. When the initial current was reduced to 1A, the total charging time increased to 122 minutes (see Figure 10.6). Table 10.1 compares the time it takes to charge the battery up to 99% SOC using three different topologies. From this table, it is clear that the CV charging topology results in the shortest charging time. However, the CV charging current is determined by the internal resistance of the battery. When the battery is close to empty, the charging current could be unsafely high (see Figure 10.4), depending on the internal resistance. The CC-CV strategy is employed to keep the initial charging current at an acceptable level. It is also clear that as the initial charging current decreases, the total charging time increases. Hence, there must be a strategy to select the initial charging current in a way that the desired objectives (defined in terms of the total charging time and energy loss) are satisfied. This problem is formally addressed in Section 10.3.

Table 10.1
Charging Time for Different Methods

Charging Method	Charging Waveform	Time to Charge
CV	Figure 10.4	90 min.
CC(1.5 A)-CV	Figure 10.5	100 min.
CC(1 A)-CV	Figure 10.6	122 min.

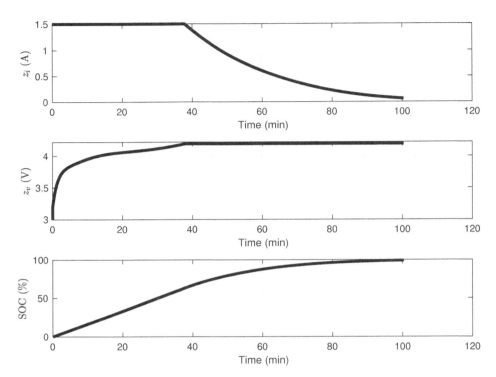

Figure 10.5 CC-CV charging. Sample current and voltage profile of CC-CV charging. The charging time is 101 minutes.

10.2.4 Multistage Constant Current Charging

Multistage constant current (MCC) charging strategy enables control of the charging current throughout the entire charging process. Here, the charging current is reduced step by step until the battery is charged to the desired SOC. Example 10.3 demonstrates a typical MCC strategy where the charging current is halved at consecutive charging stages.

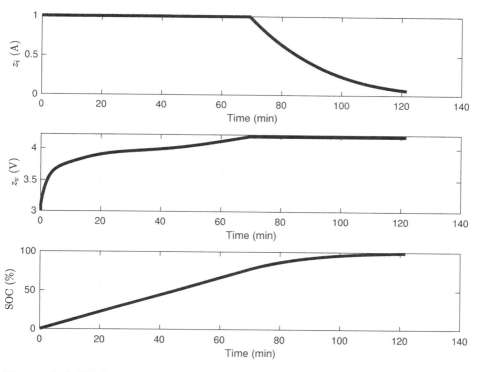

Figure 10.6 CC-CV charging. Sample current and voltage profile of CC-CV charging. The charging time is 122 minutes.

Example 10.3

Consider the battery parameters introduced in Example 10.1. This battery is charged according to the following protocol.

- Set the initial charging current to $i_c = 1.5A$.

- Use constant current charging with charging current i_c until the voltage reaches OCV_{max}.

- When the terminal voltage is $v = OCV_{max}$ reduce the current to half; for example, set $i_c = i_c/2$, and go to (a).

- Terminate charging when SOC reaches 99%.

Compute the voltage, current, and SOC throughout the charging duration and plot them. Assume that the battery is empty at the start.

Figure 10.7 shows the voltage and current of the multistage constant current charging algorithm.

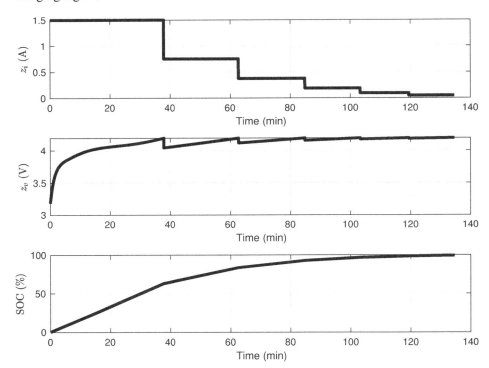

Figure 10.7 Constant current charging. Sample current and voltage profile of multistage constant charging algorithm. The total charging time is 135 minutes.

10.2.5   Pulse Charging

Pulse charging consists of multiple phases of CC or CV charging profiles with rest periods in between each phase. There are various pulse-charging strategies. Some of them are:

- Constant current constant frequency pulse charging (CCCF-PC);
- Constant current variable frequency pulse charging (CCVF-PC);

- Variable current constant frequency pulse charging (VCCF-PC).

The first group is referred to as CCCF-PC, which is shown in Figure 10.8. The other two groups are CCVF-PC and VCCF-PC. These two charging profiles are shown in Figure 10.9 and can reduce the charging time [1]. Some of the reduction of the charge time using the pulsed-charging current may be attributed to cell self-heating, which reduces cell resistance.

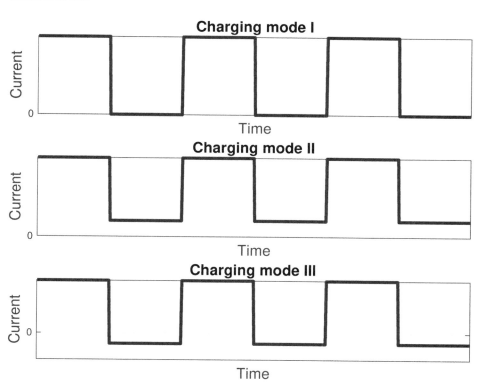

Figure 10.8 CCCF charging profile. Three different modes of constant current constant frequency pulse charging.

However, voltage pulse charging can be split into two categories. The first is based on a duty-varied voltage pulse and the second is based on a variable frequency voltage pulse. As for the duty-varied voltage pulse, it works by having a variable duty cycle

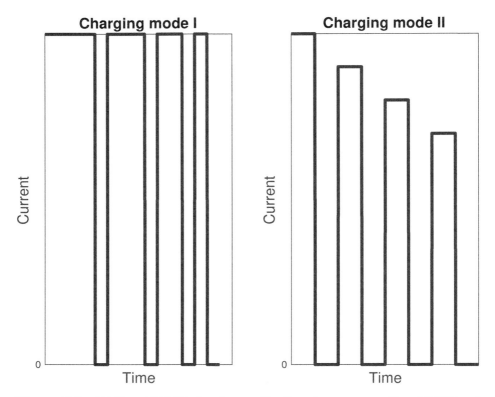

Figure 10.9 CCVF and VCCF charging profile. Sample current profile of CCCF and VCCF pulse charging.

for the voltage charging phase, while the variable frequency voltage pulse works by controlling the frequency of the voltage pulse.

10.2.6 Trickle Charging

When the battery is fully charged and left idle, some charge is lost due to self-discharge. As long as the battery is connected to a charger, the charger tries to keep the battery at a full state by continuing to replenish the lost charge, at a rate that is equal to self-discharging; this process is called trickle charging. It is important to mention that fixed current trickle charge is dangerous for Li-ion cell because it may cause increasing voltage if the fixed charging current exceeds self-discharge.

10.2.7 Float Charging

The idea of float charging is very similar to trickle charging. The difference is that the battery terminal is always kept at a certain voltage (e.g., at OCV_{max}). In other words, when a CV charger (Section 10.2.2) is never terminated at the end of charging, it starts to act as a float charger; as the SOC reduces due to self-discharge, the CV charger increases the charging current and vice versa.

10.3 OPTIMIZED CHARGING STRATEGIES

Consider the CC-CV charging topology where the battery is charged with a constant current I_{cc} until the terminal voltage reaches its maximum allowable limit OCV_{max}. After that, the charging mode is switched to constant voltage during which the charging current gradually reduces; charging is terminated when the charging current falls below a shutdown voltage I_{sd}.

The problem objective is to find the constant charging current $I_{cc} < I_{max}$ that minimizes the charging time and the energy loss. Here, I_{max} is a known upper limit provided by the battery manufacturer.

Let us assume the initial SOC to be s_0. When the battery is charged with constant I_{cc}, the charging stops when the terminal voltage v reaches OCV_{max}; let us denote the SOC of the battery at this point as s_1. Let us denote the time taken for the SOC to rise from s_0 to s_1 as t_{cc} and let us denote the OCV at this time (i.e., at the end of CC mode and at the start of CV mode) as $E_1 = E_{cc}$.

The SOC change during CC the charging is defined as

$$s_{cc} = s_1 - s_0 \qquad (10.9)$$

The CC charging time is then

$$t_{cc} = \frac{Qs_{cc}}{I_{cc}} = \frac{Q(s_1 - s_0)}{I_{cc}} \qquad (10.10)$$

The energy loss during the CC charging period is

$$EL_{cc} = I_{cc}^2 R_0 t_{cc} = I_{cc} R_0 Q(s_1 - s_0) \qquad (10.11)$$

During the CV mode, the terminal voltage is fixed at $v = \text{OCV}_{\max}$. The instantaneous current $I(t)$ varies with time and is given by

$$I(t) = \frac{\text{OCV}_{\max} - E_t}{R_0} = \frac{\text{OCV}_{\max} - f(s_t)}{R_0} \quad (10.12)$$

where E_t denotes the instantaneous OCV at time t and the SOC at time t is given by

$$s_t = s_1 + \int_{t_{cc}}^{t} \frac{I(x)}{Q} dx \quad (10.13)$$

Consider the following (Unnewehr) OCV-SOC model

$$E_k = f(s_k) = a_0 + a_1 s_k \quad (10.14)$$

for a given E_k the SOC is computed as

$$s_k = f^{-1}(E_k) = \frac{E_k - a_0}{a_1} \quad (10.15)$$

Also,

$$\text{OCV}_{\min} = a_0, \quad \text{OCV}_{\max} = a_0 + a_1 \quad (10.16)$$

Let us substitute (10.13) into (10.12)

$$\begin{aligned}
I(t) &= \frac{\text{OCV}_{\max} - E_t}{R_0} \\
&= \frac{\text{OCV}_{\max} - a_0 - a_1 \left(s_1 + \int_{t_{cc}}^{t} \frac{I(x)}{Q} dx \right)}{R_0}
\end{aligned} \quad (10.17)$$

Taking derivative on both sides, one gets

$$\frac{dI(t)}{dt} = -\frac{a_1}{QR_0} I(t) \quad (10.18)$$

By solving the above differential equation, we get

$$I(t) = \kappa e^{-(a_1/QR_0)(t - t_{cc})} \quad (10.19)$$

where constant κ can be found by substituting I_{cc} for the initial current (the initial current can be shown to be I_{cc} for the CV portion). By substituting $I(t_{cc}) = I_{cc}$, we get $\kappa = I_{cc}$. The instantaneous current is then

$$I(t) = I_{cc}e^{-(a_1/QR_0)(t-t_{cc})} \tag{10.20}$$

Let us assume that the charging is stopped when the current $I(t)$ falls below the shutdown threshold I_{sd}, that is,

$$\frac{I_{sd}}{I_{cc}} = e^{-(a_1/QR_0)t_{cv}} \tag{10.21}$$

where t_{cv} is the CV charging time given by

$$t_{cv} = t - t_{cc} = -\left(\frac{QR_0}{a_1}\right)\log\left(\frac{I_{sd}}{I_{cc}}\right) \tag{10.22}$$

The energy loss during the CV portion of the charging is then

$$\begin{aligned}
\mathrm{EL}_{cv} &= \int_0^{t_{cv}} I_{cc}^2 R_0 e^{-(2a_1/QR_0)t} dt \\
&= \frac{QI_{cc}^2 R_0^2}{2a_1}\left(1 - \frac{I_{sd}^2}{I_{cc}^2}\right) = \frac{QI_{cc}^2 R_0^2}{2a_1} - \frac{QI_{sd}^2 R_0^2}{2a_1}
\end{aligned} \tag{10.23}$$

At the point when CC ends (and CV starts), the OCV is

$$E_1 = \mathrm{OCV}_{\max} - I_{cc}R_0 \tag{10.24}$$

From this, and using the OCV model (10.15), s_1 can be written as

$$s_1 = \frac{\mathrm{OCV}_{\max} - I_{cc}R_0 - a_0}{a_1} \tag{10.25}$$

Now, the total CC-CV charging time is written as

$$t_c \triangleq t_{cccv} = t_{cc} + t_{cv} = \frac{Q(s_1 - s_0)}{I_{cc}} - \left(\frac{QR_0}{a_1}\right)\log\left(\frac{I_{sd}}{I_{cc}}\right) \tag{10.26}$$

Substituting (10.25) in t_c,

$$t_c = \frac{Q(\mathrm{OCV}_{\max} - I_{cc}R_0 - a_0 - a_1 s_0)}{a_1 I_{cc}} - \left(\frac{QR_0}{a_1}\right)\log\left(\frac{I_{sd}}{I_{cc}}\right) \tag{10.27}$$

The total energy loss is calculated as

$$
\begin{aligned}
\mathrm{EL_c} &\triangleq \mathrm{EL_{cccv}} = \mathrm{EL_{cc}} + \mathrm{EL_{cv}} \\
&= I_{cc}R_0 Q(s_1 - s_0) + \frac{Q I_{cc}^2 R_0^2}{2a_1} - \frac{Q I_{sd}^2 R_0^2}{2a_1} \\
&= I_{cc}R_0 Q\left(\frac{\mathrm{OCV_{max}} - a_0 - a_1 s_0}{a_1}\right) - \frac{Q I_{cc}^2 R_0^2}{2a_1} - \frac{Q I_{sd}^2 R_0^2}{2a_1}
\end{aligned}
\tag{10.28}
$$

The following observations can be made about the charging time t_c in (10.27) and the energy loss $\mathrm{EL_c}$ in (10.28):

1. The total CC-CV charging time t_c in (10.27) decreases with the increase of the CC charging current I_{cc}. The charging time is at its maximum when the CC charging current I_{cc} is at its minimum (i.e., when $I_{cc} = I_{min}$). Let us write the maximum charging time as

$$
t_{c,max} = t_c|_{I_{cc}=I_{min}}
\tag{10.29}
$$

2. The total CC-CV energy loss $\mathrm{EL_c}$ increases with the increase of CC charging current I_{cc}; however, it peaks when

$$
I_{cc} = I_{peak} = \frac{(\mathrm{OCV_{max}} - a_0 - a_1 s_0)}{R_0}
\tag{10.30}
$$

However, the maximum possible charging current happens when the initial SOC is $s_0 = 0$, that is,

$$
I_{max} = \frac{\mathrm{OCV_{max}} - \mathrm{OCV_{min}}}{R_0}
\tag{10.31}
$$

It is easy to see that $I_{peak} \leq I_{max}$. The maximum possible energy loss is then written as

$$
\mathrm{EL_{c,max}} = \mathrm{EL_c}|_{I_{cc}=I_{max}}
\tag{10.32}
$$

In order to unify the objective terms, the following cost functions are defined:

$$
\text{Normalized charging time cost} \quad \mathcal{J}_t = \frac{t_c}{t_{c,max}}
\tag{10.33}
$$

$$
\text{Normalized energy loss cost} \quad \mathcal{J}_e = \frac{\mathrm{EL_c}}{\mathrm{EL_{c,max}}}
\tag{10.34}
$$

Now a unified cost function is defined as follows

$$\mathcal{J} = \rho\mathcal{J}_e + (1 - \rho)\mathcal{J}_t \qquad (10.35)$$

where $\rho \in [0, 1]$ is used as a weight to combine two objective functions. The goal now is to find I_{cc} which minimizes the unified cost $\mathcal{J}$. Optimization methods, such as the Newton Raphson method, can be used to search for the optimal point with $I_{cc} = I_{min}$ as the initial value.

10.4 NUMERICAL RESULTS

In this section, a numerical demonstration of the proposed approach is presented. A typical Li-ion battery is assumed for simulation with $a_0 = \mathrm{OCV}_{min} = 3\mathrm{V}$ and $a_1 = 1.2\mathrm{V}$. This resulted in $\mathrm{OCV}_{max} = 4.2\mathrm{V}$. The battery capacity is assumed to be $Q = 1.5$ Ah and the internal resistance is assumed to be $R_0 = 0.2\,\Omega$. The initial SOC is assumed to be $s_0 = 0$. The maximum possible current for the simulation is computed to be

$$I_{max} = \frac{\mathrm{OCV}_{max} - \mathrm{OCV}_{min}}{R_0} = \frac{4.2 - 3}{0.2} = 6\mathrm{A} \qquad (10.36)$$

The simulations figures presented in this sections are computed for I_{cc} values between a selected minimum value of $I_{min} = 0.1\mathrm{A}$ and $I_{max} = 6\mathrm{A}$.

Figure 10.10(a) shows the total energy loss EL_c in (10.28) and the total charging time t_c in (10.27) for different values of I_{cc}. The energy loss EL_c is computed to Joules (Watt seconds) by multiplying EL_c by 3600. It can be noticed in Figure 10.10(a) that the energy loss has a monotonously increasing trend with the charging current for the entire values of $I_{cc} \in [I_{min}, I_{max}]$.

Figure 10.10(b) shows an extended version of Figure 10.10(a) for longer values of I_{cc}. This extended figure is meant to show that the peak energy loss happens when $I_{cc} = I_{max}$.

Figure 10.11 shows the normalized cost functions $\mathcal{J}_t$ and $\mathcal{J}_e$ for all possible values of the CC charging current I_{cc}. It shows the charging time and energy loss as monotonically decreasing and increasing functions, respectively, of I_{cc}. This also explains the competing objectives of fast charging (i.e., reducing the charging time means a higher cost in terms of energy loss).

Figure 10.12 shows the unified cost function for different combinations of weights. The cost $\mathcal{J}$ corresponding to $\rho = 0.5$ shows the scenario where charging time and energy loss are given equal priority. The cost $\mathcal{J}$ corresponding to $\rho = 0.9$ shows a

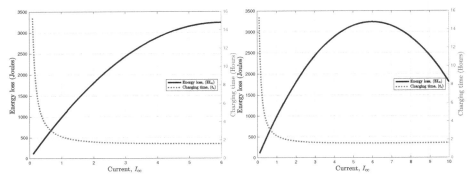

(a) Total energy loss and charging time for possible (b) Total energy loss and charging time for hypotheti-
CC charging currents.   cal CC charging with extended range for I_{cc}.

Figure 10.10 Total energy loss and charging time

scenario where energy loss is prioritized over charging time; for this case, the charging current is lower (and the charging time is higher) compared to an equal weight scenario.

10.5 SUMMARY

Li-ion battery charges are limited in terms of how much current and voltage they can apply during charging. Applying high voltage/current, within the allowable limits, may reduce charging time at the cost of high energy loss and resulting temperature increase. In this chapter, an optimized strategy was demonstrated to determine a balanced CC-CV charging approach considering both charging time and energy loss constraints. It is also known that higher charging voltage and current cause a state of health decay in batteries. The knowledge about optimized charging under SOH constraints is still in the early stages; it is an active research topic. Some of the pulse-charging strategies discussed in this chapter are considered candidates for optimized charging under SOH constraints.

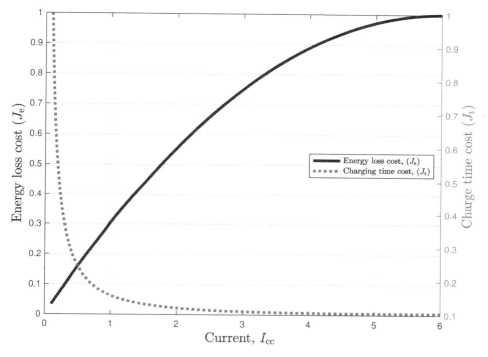

Figure 10.11 Normalized cost functions.

10.6 BIBLIOGRAPHICAL NOTES

In [2], an early version of the smart charging work is presented. A detailed literature review of existing smart charging approaches can be found in [1]. Some discussions about optimal charging under SOH constraints can be found in [2].

References

[1] Q. Lin, J. Wang, R. Xiong, W. Shen, and H. He, "Towards a smarter battery management system: A critical review on optimal charging methods of lithium ion batteries," *Energies*, Vol. 183, pp. 220–234, 2019.

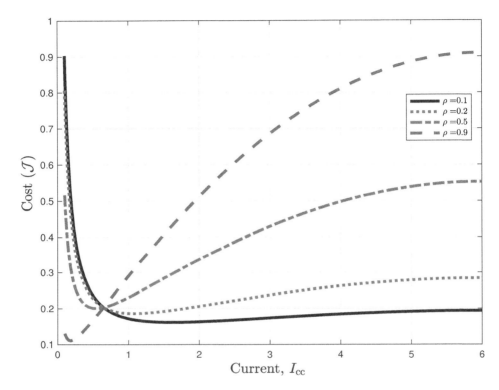

Figure 10.12 Unified cost function.

[2] A. Abdollahi, X. Han, N. Raghunathan, B. Pattipati, B. Balasingam, K. R. Pattipati, Y. Bar-Shalom, and B. Card, "Optimal charging for general equivalent electrical battery model, and battery life management," *Journal of Energy Storage*, Vol. 9, pp. 47-58, 2017.

Chapter 11

Evaluation and Benchmarking of Battery Management Systems

11.1 INTRODUCTION

We have seen that a battery management system performs various operations, such as battery fuel gauging, optimized charging, cell balancing, and thermal balancing. How do we tell that a BMS is working as it is supposed to? In other words, how do we evaluate the performance of a BMS?

To give an analogy, let us consider a bathroom scale. How do we know that the scale displays the correct weight when we step on it? In the case of household scales, they were likely calibrated or checked before being sold to the consumers. However, it is hard to validate their accuracy once they are in a household. Unless they go into complete malfunction, household customers have no way of knowing the accuracy of a scale. In a laboratory setting, a bathroom scale can be easily tested. For example, they can be used to measure the weights of objects whose weight is precisely known, or they can be applied with an industry-scale force that is accurate to micro-Newtons and the corresponding measurements of the scale can be recorded to check for its accuracy.

Evaluating the performance of a battery management system is a challenging and less-studied problem to date. Unlike the availability of high-precision scales in the previous example, there are no high-precision measurement systems that can directly measure the SOC of a battery. The precise computation of SOC requires the knowledge of many variables, such as battery capacity, ECM parameters, and hysteresis voltage. High-precision SOC estimation starts with high-precision voltage and current measurements followed by careful modeling and parameter estimation approaches that eventually lead to high-precision SOC measurements that can be compared against the SOC estimates offered by a commercial BFG device. This chapter explains the details of such an approach to validate the SOC and TTS estimates of a battery fuel gauge based on the following three different metrics: the Coulomb counting metric, the OCV metric, and the time-to-voltage (TTV) metric.

11.2 COULOMB COUNTING METRIC

The Coulomb counting approach [1] allows one to recursively compute the SOC at time instant k as follows:

$$s_{cc}(k) = s_{cc}(k-1) + \frac{\Delta_k i(k)}{3,600Q} \tag{11.1}$$

where $s_{cc}(k)$ denotes the SOC at time k. Q denotes the battery capacity in Ah, $s_{cc}(0)$ denotes the initial SOC, Δ_k is the sampling time in seconds, and $i(k)$ is the current through the battery ($i(k) > 0$ denotes charging; and $i(k) < 0$ denotes discharging). The above SOC computation approach is quite accurate for short time durations [1] when the initial SOC $s_{cc}(0)$ and the battery capacity Q are perfectly known.

Let us denote the SOC estimate of a BFG at time k as $s_{bfg}(k)$. The instantaneous Coulomb counting error can be written as [1]

$$\epsilon_{cc}(k) = s_{cc}(k) - s_{bfg}(k) \tag{11.2}$$

11.3 OCV-SOC METRIC

The OCV-SOC characterization of the battery gives a look-up procedure to estimate the SOC. Let us denote the OCV measurement on a rested battery as $v_r(k)$. Given the parameters of the OCV-SOC function $f(\cdot)$, the SOC can be estimated using the inverse lookup

$$s_{ocv}(k) = f_o^{-1}(v_r(k)) \tag{11.3}$$

The OCV-SOC error is then computed as

$$\epsilon_{ocv} = s_{bfg}(k) - s_{ocv}(k) \tag{11.4}$$

11.4   TTV METRIC

The TTV metric is the error between the time predicted by the BFG for the terminal voltage to reach a certain value and the actual time taken for it. Figure 11.1 explains the procedure to compute the TTV metric.

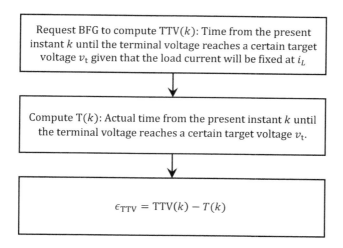

Figure 11.1 Procedure to compute the TTV metric.

Most commercially available BFGs may not have TTV as an output that can be set to any terminal voltage. Rather, many of them will have the predicted time until the battery needs to shut down. In that case, the TTV metric can be modified to compute the TTS metric. In other words, the TTS metric is a special case of the TTV metric.

Remark 11.1 In an electric vehicle, the TTS is converted to remaining mileage and continuously reported to the end user on the dashboard. In the absence of TTV (or TTS), a similar metric can be created based on the reported remaining mileage.

When a commercially available BFG does not have either TTV or TTS output, it may be computed. Based on the parameters introduced in Table 11.1, computing TTV involves the following steps:

1. Compute the voltage drop v_d.

The voltage drop accounts for the hysteresis and relaxation effects within the battery. In a simple R-int model, the voltage drop could be computed as

$$v_d = i_d R_0(\text{SOC}, \text{T}) \tag{11.5}$$

where i_d is the discharge current and $R_0(\text{SOC}, \text{T})$ is the internal resistant of the battery, which is indicated to be a function fo the SOC and temperature. It must be noted that the discharge current needs to be a constant so that v_d can be expected to remain constant; under the assumptions that the resistance also remains constant during the test.

2. Compute target OCV

$$V_{ot} = v_t - v_d \tag{11.6}$$

3. Compute target SOC

$$s_t = f_o^{-1}(V_{ot}) \tag{11.7}$$

4. Compute SOC difference

$$s_d(k) = s(k) - s_t \tag{11.8}$$

5. Compute TTV

$$\text{TTV}(k) = \frac{s_d(k)Q}{i_L} \tag{11.9}$$

Remark 11.2 It is important to realize that the five-step process explained above is not part of the BFG evaluation process. However, in the absence of TTV/TTS or remaining mileage output from the BFG, the above procedure can be used to compute the TTV.

Remark 11.3 If the TTV needs to be computed on behalf of the BFG (when TTV/TTS or remaining mileage is not available), any error made in computing the TTV will conflate the resulting TTV error. Hence, it is important to make sure all the necessary precautions are taken to reduce error when computing TTV. Particularly, temperature and SOC-dependent factors must be taken into consideration when computing the voltage drop v_d in (11.6).

Table 11.1

Variables Involved in TTV Computation

Quantity	Notation
Present time index	k
Present terminal voltage	$v(k)$
Present OCV	$V_o(k)$
Present SOC	$s(k)$
Target terminal voltage	v_t
Target OCV	V_{ot}
Target SOC	s_t
Voltage drop	v_d
Constant load current	i_L

Figure 11.2 shows the parameters required for TTV computation with reference to an OCV-SOC curve and the (measured) terminal voltage curve with respect to the SOC of the battery.

Let us assume that the present SOC is $s(k)$ and that the present terminal voltage of the battery is $v(k)$ while the load current is i_L. Given that the load current i_L is to remain constant, the time to reach the target voltage v_t can be computed according to (11.9). Based on that, the TTV metric is computed as follows:

$$\epsilon_{ttv} = \text{TTV}(k) - T(k) \tag{11.10}$$

where $T(k)$ is the measured time that it took from the present time epoch k until the target voltage v_t is reached at the terminals.

The TTV metric is computed based on an independent and accurate quantity, measured time. In order to compute the TTV metric, the following additional state and parameter estimates from the BFG are needed: SOC $s(k)$, battery capacity Q, OCV-SOC curve $f(\cdot)$, and the voltage drop v_d. It is the BFG that needs to keep all these state and parameter estimates accurate. Hence, the TTV metric should be considered a measure of the overall accuracy of the BFG rather than just the SOC estimation error. For example, incorrectly computed ECM parameters will result in incorrect voltage drop computations, leading to incorrect TTV computations. Such a scenario is depicted in Figure 11.3.

The TTV prediction error is affected by the following factors:

1. The magnitude of the current i_L. The TTV in (11.9) is inversely proportional to the load current i_L.

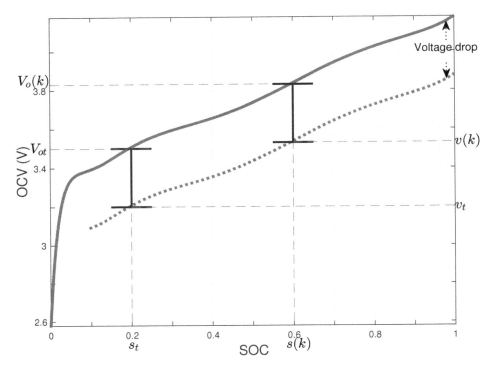

Figure 11.2   Variables involved in TTV computation. Reproduced with permission from [2].

2. The SOC difference $s_d(k)$ in (11.9) is affected by the difference between the present voltage $v(k)$ and the target voltage v_t.

3. Present OCV $V_o(k)$. The TTV in (11.9) is affected by the slope of the OCV-SOC curve.

11.5   DEMONSTRATION OF THE BFG EVALUATION

This section presents a detailed approach to BFG evaluation using the battery simulator introduced in Chapter 3. First, a load current profile is created specifically for BFG evaluation. This load current profile is used to simulate voltage across the battery

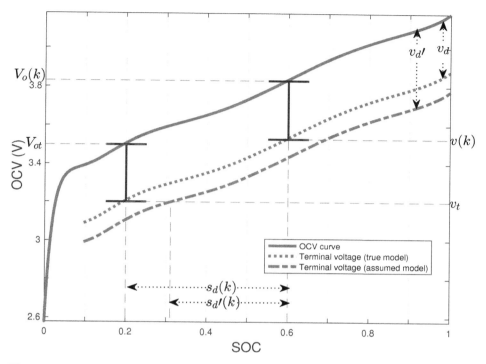

Figure 11.3 Variables involved in TTV error computation. Reproduced with permission from [2].

terminals. Based on this voltage and current data, the SOC of the battery and the TTV for a given target voltage v_t is computed. The three evaluation metrics described in this chapter are computed based on the computed SOC and TTV.

In a real-world setting, a similar current profile needs to be applied to the battery. The BFG under evaluation need to compute the SOC and TTV based on the measured voltage and current through the battery. The remaining steps of a realistic BFG evaluation will be very similar to what is described next in this section.

Figure 11.4 shows the BFG evaluation current profile along with the computed voltage using the battery simulator. In order to perform BFG evaluation using this model, the battery capacity and the OCV-SOC model needs to be perfectly known. These two quantities can be obtained by an OCV characterization (see details in Chapter 4) just prior to the BFG evaluation procedure. The 9-hour BFG evaluation load profile, shown in Figure 11.4, consists of the following components:

- A 0.3A discharge for 1 hour to bring the SOC down followed by 1 hour of rest.

- A 2-hour dynamic load. The BFG is started at an arbitrary point during the dynamic load.

- An hour of rest. The true SOC remains the same during the rest. After allowing sufficient time to rest the battery, the OCV-lookup error is computed towards the end of this rest period.

- A 0.3A discharge for 5 hours. The TTV metric is computed during this constant-current discharge.

The above BFG evaluation load profile is applied to the battery simulator and the terminal voltage $v(k)$ is obtained. The plot at the bottom of Figure 11.4 shows the computed terminal voltage according to the R-int model. Starting from an arbitrary time, the voltage and current values are fed to the EKF-based battery fuel gauge detailed in Chapter 8 to estimate the SOC $s_{bfg}(k)$. The starting point of the BFG in Figure 11.4 is marked as "Start BFG" by the 3-hour mark. The TTV is computed at the 6-hour mark using (11.9) for a target voltage of $v_t = 3.5V$.

In a realistic BMS evaluation, a similar current profile to the one shown in Figure 11.4 will be applied to the battery and the SOC and TTS reported by the BFG are recorded. The TTS values reported by a BFG are a special case of TTV computed when the target voltage is $v_t = OCV_{min}$. Computing TTV at various target voltages, other than the shutdown voltage, makes the evaluation more robust.

The first row of Table 11.2 shows the following quantities that need to be computed for BFG evaluation.

- The SOC estimate is given by the BFG $s_{bfg}(k_1)$ at a certain point in time. In the present example, the SOC is computed at time instant k_1 corresponding to the 4.5-hour mark indicated on the BFG evaluation load profile in Figure 11.4. In a generic BFG evaluation, any time instant can be selected.

- The true SOC estimate is based on the OCV-lookup method when the battery is at rest $s_{ocv}(k_1)$. The OCV-based SOC estimate can only be computed when the battery is rested (i.e., when the measured voltage is the same as the OCV). It is possible to compute OCV by subtracting the voltage drop from the terminal voltage; however, this approach is avoided due to the uncertainties involved in voltage drop computation.

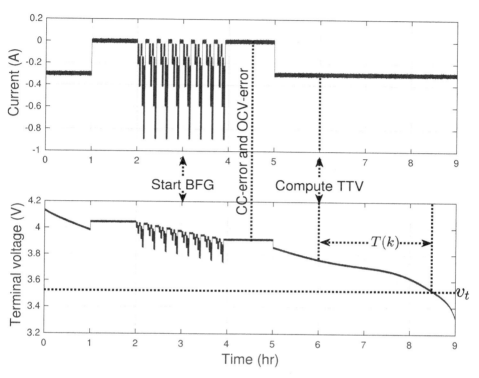

Figure 11.4 Details of the BFG evaluation. Reproduced with permission from [2].

- The true SOC estimate is based on the Coulomb counting approach $s_{cc}(k_2)$. In order to compute $s_{cc}(k_2)$, the battery capacity and the initial SOC are needed. For the present approach, the battery capacity must be computed just before the BFG evaluation; usually an OCV test precedes the BFG evaluation load profile. The OCV-test allows computing both the capacity and the OCV-SOC model parameters accurately for the evaluation.

- The time to voltage is computed by the BFG $TTV(k_2)$ starting from the time instant k_2 corresponding to the 6-hour mark indicated on the BFG evaluation load profile in Figure 11.4 until the terminal voltage reached $v_t = 3.5V$.

- The true time to voltage is computed by waiting until the target voltage $v_t = 3.5V$ is reached at the battery terminals.

Table 11.2

BFG Evaluation Data for 10 Independent Experiments

	SOC(BFG) $s_{\text{bfg}}(k_1)$	SOC(CC) $s_{\text{cc}}(k_1)$	SOC(OCV) $s_{\text{ocv}}(k_1)$	TTV (true) $T(k_2)$	TTV(BFG) $\text{TTV}(k_2)$
1	0.9985	0.9961	0.6769	2.7184	2.6242
2	0.9976	0.9998	0.6821	2.6816	2.5449
3	0.9986	0.9998	0.6923	2.6592	2.6143
4	0.9994	0.9998	0.6658	2.6369	2.6282
5	0.9988	0.9998	0.6894	2.6383	2.6648
6	0.9985	0.9998	0.6796	2.6263	2.5559
7	0.9990	0.9891	0.6880	2.5847	2.5525
8	0.9991	0.9998	0.6263	2.6053	2.6218
9	0.9992	0.9998	0.6468	2.6602	2.5897
10	0.9987	0.9998	0.6628	2.6236	2.6133

The values shown in the first row of Table 11.2 are obtained for a particular Monte Carlo run. The experiment is repeated on several independent Monte Carlo runs. Each row in Table 11.2 provides the recorded parameters from 10 Monte Carlo runs. Using the evaluation data from each row of Table 11.2, the following BFG evaluation errors are computed: the Coulomb counting error is computed according to (11.2), the OCV error is computed according to (11.4), and the TTV error is computed according to (11.10). Table 11.3 summarizes these errors.

Next, we discuss an approach to combine the BFG evaluation errors, shown in Table 11.3. For this, the following mean square root error (MSRE) is introduced:

$$\text{MSRE}(\epsilon_{\text{cc}}^i(k_1)) \triangleq \text{MSRE}(\epsilon_{\text{cc}}) = \sqrt{\sum_{i=1}^{N_r}(\epsilon_{\text{cc}}^i(k_1))^2} \qquad (11.11)$$

$$\text{MSRE}(\epsilon_{\text{occ}}^i(k_1)) \triangleq \text{MSRE}(\epsilon_{\text{ocv}}) = \sqrt{\sum_{i=1}^{N_r}(\epsilon_{\text{ocv}}^i(k_1))^2} \qquad (11.12)$$

$$\text{MSRE}(\epsilon_{\text{ttv}}^i(k_2)) \triangleq \text{MSRE}(\epsilon_{\text{ttv}}) = \sqrt{\sum_{i=1}^{N_r}(\epsilon_{\text{ttv}}^i(k_2))^2} \qquad (11.13)$$

Table 11.3

BFG Evaluation Error for 10 Independent Experiments

	CC error $\epsilon_{cc}(k_1)$	OCV error $\epsilon_{ocv}(k_1)$	TTV error $\epsilon_{ttv}(k_2)$
1	0.3192	-0.0259	0.0942
2	0.3177	-0.0270	0.1366
3	0.3075	-0.0308	0.0450
4	0.3340	-0.0325	0.0087
5	0.3104	-0.0287	-0.0265
6	0.3202	-0.0315	0.0704
7	0.3010	-0.0288	0.0322
8	0.3735	-0.0286	-0.0165
9	0.3530	-0.0273	0.0705
10	0.3370	-0.0302	0.0103

where $\epsilon_{cc}^i(k_1)$ is the Coulomb counting error (11.2) computed for a specific run i and N_r is the number of such runs. Similarly, $\epsilon_{ocv}^i(k_1)$ and $\epsilon_{ttv}^i(k_1)$ are the OCV lookup error and TTV error, respectively, for specific run i.

The first row of Table 11.4 presents the MSRE values computed for $N_r = 100$ Monte Carlo runs for a particular measurement noise scenario where the voltage and current measurement noise standard deviations were selected to be $\sigma_i = 2A$ and $\sigma_v = 2V$, respectively. The remaining four rows of Table 11.4 show the MSRE values for increasing amounts of noise.

Table 11.4

BFG Evaluation Error for 10 Independent Experiments

σ	MSRE(ϵ_{cc})	MSRE(ϵ_{ocv})	MSRE(ϵ_{ttv})
2	0.3335	0.0067	1.5287
4	0.3327	0.0226	4.4249
6	0.3312	0.0331	9.7635
8	0.3310	0.0153	9.4793
10	0.3294	0.0082	14.9027

The evaluation results presented in Table 11.4 indicate a very accurate BFG that produces less than 1% in SOC error and a very small TTV error. The low error values are due to the fact that the BFG algorithm is applied to data generated from a battery simulator that employed a very simplified R-int model; the ECM parameters of the battery model are assumed perfectly known; the battery capacity is perfectly known; the OCV model parameters are perfectly known; and no hysteresis was introduced during battery simulation. There will be uncertainties with each of these assumptions during realistic BFG evaluation, as a result, the MSRE values will be high. The scope of this chapter is limited to introducing BFG evaluation metrics and explaining how to implement them. Demonstrating a real-world BFG is left as an exercise. The reader will find the details of a real-world BFG evaluation in [3] which used similar metrics presented in this chapter.

11.6 SUMMARY

This chapter discusses the importance of BMS evaluation and outlines some approaches to evaluating a BFG. Three BFG evaluation metrics are detailed and their implementation details are elaborated. It must be stressed that BFG is the most important part of a BMS and hence, BFG evaluation is very important for performance monitoring. However, additional rigorous approaches and standards must be developed to broaden BMS evaluation. Evaluation methods for other blocks of a BMS (optimal charging algorithm, cell balancing circuitry, and battery thermal management system) need to be developed as well.

11.7 BIBLIOGRAPHICAL NOTES

The three BFG evaluation metrics proposed in this chapter are based on [3]. Compared to [3], some modifications were made for each metric in this chapter to ensure the robustness of the evaluation.

References

[1] K. Movassagh, A. Raihan, B. Balasingam, and K. Pattipati, "A critical look at coulomb counting approach for state of charge estimation in batteries," *Energies*, 2021.

[2] P. Pillai, and B. Balasingam, "Approach for Rigorous Evaluation of a Battery Fuel Gauge," *IEEE Electrical Power and Energy Conference (EPEC)*, pp. 176–181, 2022.

[3] G.V. Avvari, B. Pattipati, B. Balasingam, K.R. Pattipati, and Y. Bar-Shalom, "Experimental set-up and procedures to test and validate battery fuel gauge algorithms," *Applied Energy*, Vol. 160, pp. 404–418, 2015.

Appendix A

Closed-Form Derivation of the TLS Estimate

Let us write the 2×2 matrix $\mathbf{A} \triangleq \mathbf{S}_{\mathbf{H}}^{\kappa}$ as

$$\mathbf{A} = \begin{bmatrix} \sigma_{11} & \sigma_{12} \\ \sigma_{12} & \sigma_{22} \end{bmatrix} \tag{A.1}$$

The eigenvalues of $\mathbf{A}$ can be written as

$$\lambda_1 = \frac{\sigma_{11} + \sigma_{22} + \sqrt{(\sigma_{11} - \sigma_{22})^2 + 4(\sigma_{12})^2}}{2} \tag{A.2}$$

$$\lambda_2 = \frac{\sigma_{11} + \sigma_{22} - \sqrt{(\sigma_{11} - \sigma_{22})^2 + 4(\sigma_{12})^2}}{2} \tag{A.3}$$

where λ_1 is the largest eigenvalue and λ_2 is the smallest eigenvalue. The eigenvector corresponding to λ_2 is

$$\mathbf{v}_2^{\kappa} = \begin{bmatrix} \frac{-\sigma_{12}}{\sqrt{\sigma_{12}^2 + (\sigma_{11} - \lambda_2)^2}} \\ \frac{\sigma_{11} - \lambda_2}{\sqrt{\sigma_{12}^2 + (\sigma_{11} - \lambda_2)^2}} \end{bmatrix} \tag{A.4}$$

The estimated capacity now becomes

$$\begin{aligned} \hat{Q}_{\text{TLS}}^{-1}[\kappa] &= -\frac{\mathbf{v}_2^{\kappa}(1)}{\mathbf{v}_2^{\kappa}(2)} \\ &= \frac{\mathbf{S}_{\mathbf{H}}^{\kappa}(1,2)}{\mathbf{S}_{\mathbf{H}}^{\kappa}(1,1) - \Lambda^{\kappa}(2,2)} \end{aligned} \tag{A.5}$$

Appendix B

Formal Derivation of Capacity

In Chapter 7, the battery capacity was derived in the inverse form, that is, instead of the capacity Q, the inverse capacity $1/Q$ was estimated. In this appendix, a formal approach is provided to convert the inverse capacity estimates into capacity estimates.

B.1   TRANSFORMATION OF THE INVERSE ESTIMATES

In this appendix, we present an approach to get the capacity estimate and the estimation error variance based on the inverse estimate and the inverse estimation error variance. Our derivations are based on [1, 2]. Let us assign simple variables for the inverse capacity estimate and the error variance; that is,

$$x \triangleq Q_{\mathrm{TLS}}^{-1}[\kappa] \text{ or } Q_{\mathrm{LS}}^{-1}[\kappa] \tag{B.1}$$

$$x_0 \triangleq E\{x\} = \hat{Q}_{\mathrm{TLS}}^{-1}[\kappa] \tag{B.2}$$

$$P_x \triangleq E\{(x - x_0)^2\} = P_{\mathrm{TLS}}[\kappa] \tag{B.3}$$

Defining

$$y \triangleq f(x) = \frac{1}{x} \tag{B.4}$$

our objective is to find approximations for $E\{y\}$ and $E\{(y - E\{y\})^2\}$.

277

B.1.1   The Expected Value of y

The second-order Taylor series approximation is given by

$$y = f(x) = f(x_0) + f'(x_0)(x - x_0) + \frac{1}{2}f''(x_0)(x - x_0)^2 \tag{B.5}$$

The second order approximation of $E\{y\}$ is given by

$$E\{y\} = E\{f(x_0)\} + f'(x_0)\underbrace{\left(E\{x\} - E\{x_0\}\right)}_{=0} + \frac{1}{2}f''(x_0)E\{(x - x_0)^2\}$$

$$= \frac{1}{x_0} + \frac{P_x}{x_0^3} \tag{B.6}$$

B.1.2   The Variance of the Expected Value of y

Let us expand $f(x)$ as a first-order Taylor series around the true value x_0.

$$y = f(x) = f(x_0) + f'(x_0)(x - x_0) \tag{B.7}$$

The variance of y is given by

$$E\{(y - E\{y\})^2\} = E\{(f'(x_0)(x - x_0))^2\} = \frac{P_x}{x_0^4} \tag{B.8}$$

Now the expected values of the capacity estimate and its estimation error variance are given by

$$\hat{Q}_{\text{TLS}}[\kappa] = \frac{1}{\hat{Q}_{\text{TLS}}^{-1}[\kappa]} + \frac{P_{\text{TLS}}[\kappa]}{\left(\hat{Q}_{\text{TLS}}^{-1}[\kappa]\right)^3} \tag{B.9}$$

$$R_{\text{TLS}}[\kappa] = \frac{P_{\text{TLS}}[\kappa]}{\left(\hat{Q}_{\text{TLS}}^{-1}[\kappa]\right)^4} \tag{B.10}$$

References

[1] R.C. Elandt-Johnson, and N.L. Johnson, *Survival models and data analysis*, John, Wiley & Sons, New York, 1980.

[2] A. Stuart, and J.K. Ord, *Kendall's Advanced Theory of Statistics, Distribution Theory (Vol. 1)*, John Wiley & Sons. Arnold, London, 2010.

Appendix C

Discretization of the State-Space Model

Consider the following continuous time-state space model

$$\dot{x}(t) = Ax(t) + Bu(t) \qquad \text{(C.1)}$$

The discretized system can be given by

$$x[k+1] = Gx[k] + Hu[k] \qquad \text{(C.2)}$$

where

$$G = e^{A\Delta t}, \quad H = \left(\int_0^{\Delta t} e^{A\lambda} d\lambda \right) B \qquad \text{(C.3)}$$

and Δt is the sampling time.

List of Acronyms

AC	Alternating current
AR-ECM	Adaptive Randles ECM
BEV	Battery electric vehicles
BFG	Battery fuel gauge
BMS	Battery management systems
BTMS	Battery thermal management system
CC	Constant current
CC-CV	Constant current-constant voltage
CV	Constant voltage
CF	Capacity fade
CRLB	Cramer-Rao lower bound
CT	Charge transfer
DC	Direct current
DL	Double layer
ECM	Equivalent circuit model
EIS	Electroimpedance spectroscopy

EV	Electric vehicle
EKF	Extended Kalman filter
EMF	Electromotive force
KF	Kalman filter
LS	Least square
MCC	Multistage constant current
MSRE	Mean square root error
NIS	Normalized innovation squared
NMSE	Normalized mean square error
OCV	Open circuit voltage
PCM	Phase change materials
PEG	Polyethylene glycol
PF	Power fade
RUL	Remaining useful life
SEI	Solid electrolyte interface
SNR	Signal to noise ratio
SOC	State of charge
SOH	State of health
TECM	Thermal-electric equivalent circuit model
TTS	Time to shutdown
TTV	Time to voltage

About the Author

Balakumar Balasingam is an associate professor in the Department of Electrical and Computer Engineering at the University of Windsor. Before that, he was an assistant research professor in the Department of Electrical and Computer Engineering at the University of Connecticut. He received his PhD in electrical engineering from McMaster University, Canada, in 2008. After that, he held two postdoctoral positions, one at the University of Ottawa from 2008 to 2010 and another at the University of Connecticut from 2010 to 2012. Dr. Balasingam founded the Battery Management Research Lab (BMSLab, www.bmslab.org) at the University of Windsor in 2017. Through the BMSLab, Dr. Balasingam has been collaborating with industry, academic, and community partners toward research and educational activities involving battery management systems.

Index

Telecommunication Networks for the Smart Grid, Alberto Sendin, Miguel A. Sanchez-Fornie, Iñigo Berganza, Javier Simon, and Iker Urrutia

A Whole-System Approach to High-Performance Green Buildings, David Strong and Victoria Burrows

For further information on these and other Artech House titles, including previously considered out-of-print books now available through our In-Print-Forever® (IPF®) program, contact:

Artech House
685 Canton Street
Norwood, MA 02062
Phone: 781-769-9750
Fax: 781-769-6334
e-mail: artech@artechhouse.com

Artech House
16 Sussex Street
London SW1V 4RW UK
Phone: +44 (0)20 7596-8750
Fax: +44 (0)20 7630-0166
e-mail: artech-uk@artechhouse.com

Find us on the World Wide Web at: www.artechhouse.com